AF347531

Marine Resources Application Potential for Biotechnological Purposes

Marine Resources Application Potential for Biotechnological Purposes

Editors

Marco F. L. Lemos
Sara C. Novais
Susana F. J. Silva
Carina Félix

MDPI • Basel • Beijing • Wuhan • Barcelona • Belgrade • Manchester • Tokyo • Cluj • Tianjin

Editors
Marco F. L. Lemos
Marine and Environmental
Sciences Centre (MARE)
Polytechnic of Leiria
Peniche
Portugal

Sara C. Novais
Marine and Environmental
Sciences Centre (MARE)
Polytechnic of Leiria
Peniche
Portugal

Susana F. J. Silva
Marine and Environmental
Sciences Centre (MARE)
Polytechnic of Leiria
Peniche
Portugal

Carina Félix
Marine and Environmental
Sciences Centre (MARE)
Polytechnic of Leiria
Peniche
Portugal

Editorial Office
MDPI
St. Alban-Anlage 66
4052 Basel, Switzerland

This is a reprint of articles from the Special Issue published online in the open access journal *Applied Sciences* (ISSN 2076-3417) (available at: www.mdpi.com/journal/applsci/special_issues/ marine_resources_biotechnological).

For citation purposes, cite each article independently as indicated on the article page online and as indicated below:

LastName, A.A.; LastName, B.B.; LastName, C.C. Article Title. *Journal Name* **Year**, *Volume Number*, Page Range.

ISBN 978-3-0365-1736-0 (Hbk)
ISBN 978-3-0365-1735-3 (PDF)

Cover image courtesy of Nuno Vasco Rodrigues

Contents

About the Editors . . . . **vii**

Marco F. L. Lemos, Sara C. Novais, Susana F. J. Silva and Carina Félix
Marine Resources Application Potential for Biotechnological Purposes
Reprinted from: *Applied Sciences* **2021**, *11*, 6074, doi:10.3390/app11136074 . . . . **1**

Sushanta Kumar Saha, Hande Ermis and Patrick Murray
Marine Microalgae for Potential Lutein Production
Reprinted from: *Applied Sciences* **2020**, *10*, 6457, doi:10.3390/app10186457 . . . . **5**

Bernardo Duarte, Eduardo Feijão, Johannes W. Goessling, Isabel Caçador and Ana Rita Matos
Pigment and Fatty Acid Production under Different Light Qualities in the Diatom *Phaeodactylum tricornutum*
Reprinted from: *Applied Sciences* **2021**, *11*, 2550, doi:10.3390/app11062550 . . . . **23**

Mari Carmen Ruiz-Domínguez, Pedro Cerezal, Francisca Salinas, Elena Medina and Gabriel Renato-Castro
Application of Box-Behnken Design and Desirability Function for Green Prospection of Bioactive Compounds from *Isochrysis galbana*
Reprinted from: *Applied Sciences* **2020**, *10*, 2789, doi:10.3390/app10082789 . . . . **43**

Natalie L. Pino-Maureira, Rodrigo R. González-Saldía, Alejandro Capdeville and Benjamín Srain
Rhodotorula Strains Isolated from Seawater That Can Biotransform Raw Glycerol into Docosahexaenoic Acid (DHA) and Carotenoids for Animal Nutrition
Reprinted from: *Applied Sciences* **2021**, *11*, 2824, doi:10.3390/app11062824 . . . . **61**

Rafael Félix, Ana M. Carmona, Carina Félix, Sara C. Novais and Marco F. L. Lemos
Industry-Friendly Hydroethanolic Extraction Protocols for *Grateloupia turuturu* UV-Shielding and Antioxidant Compounds
Reprinted from: *Applied Sciences* **2020**, *10*, 5304, doi:10.3390/app10155304 . . . . **77**

Carina Félix, Rafael Félix, Ana M. Carmona, Adriana P. Januário, Pedro D.M. Dias, Tânia F.L. Vicente, Joana Silva, Celso Alves, Rui Pedrosa, Sara C. Novais and Marco F.L. Lemos
Cosmeceutical Potential of *Grateloupia turuturu*: Using Low-Cost Extraction Methodologies to Obtain Added-Value Extracts
Reprinted from: *Applied Sciences* **2021**, *11*, 1650, doi:10.3390/app11041650 . . . . **97**

Glacio Souza Araujo, João Cotas, Tiago Morais, Adriana Leandro, Sara García-Poza, Ana M. M. Gonçalves and Leonel Pereira
Calliblepharis jubata Cultivation Potential—A Comparative Study between Controlled and Semi-Controlled Aquaculture
Reprinted from: *Applied Sciences* **2020**, *10*, 7553, doi:10.3390/app10217553 . . . . **113**

Lihua Xu, Dengfeng Li, Yigang Tong, Jing Fang, Rui Yang, Weinan Qin, Wei Lin, Lingtin Pan and Wencai Liu
A Novel Singleton Giant Phage Yong-XC31 Lytic to the *Pyropia* Pathogen *Vibrio mediterranei*
Reprinted from: *Applied Sciences* **2021**, *11*, 1602, doi:10.3390/app11041602 . . . . **125**

About the Editors

Marco F. L. Lemos

Marco Lemos (Ph.D.) is an associate professor at the Polytechnic of Leiria and researcher at the Marine and Environmental Sciences Centre (MARE) focusing his research on the study of ecotoxicological modes of action and also in the marine biotechnology field. Currently he is the Director of the COVID-19 Diagnostic Centre of the Polytechnic of Leiria. He was the Coordinator of R&D Units Marine and Environmental Sciences Centre (MARE-IPLeiria) and Marine Resources Research Group (GIRM), and the Coordinator of CETEMARES building (Marine Sciences R&D, Education, and Knowledge Dissemination Centre), and also the national MARE Biotechnology line coordinator. He is the PI of several national and international projects including other research centres and the industry.

Sara C. Novais

Sara Novais (Ph.D.) holds a Ph.D. in Biology and is currently an assistant researcher at the Polytechnic of Leiria and coordinator of the Biotechnology and Resources Valorisation research line of MARE. She was part of the coordination board of MARE-IPLeria. Her scientific interests are related to the study of toxicological mechanisms, pathways and chemical modes of action using technologies and other biomarker tools, applied to the fields of ecotoxicology, ecology, aquaculture, and biotechnology. She is involved in several national and international projects with both academia and industry, supervising several Ph.D. and M.Sc. students and actively publishing in the field.

Susana F. J. Silva

Susana Silva (Ph.D.) holds a Ph.D. in Food Sciences (University of Reading, UK, 2008) and is now an associate professor at the Polytechnic of Leiria and member of MARE-PLeiria. Has been involved and coordinated several projects with food chain stakeholders involvement, which have resulted in several published findings, novel food product prototypes and patented processes and food packaging solutions. Her current main research goal is to actively contribute towards more sustainable food products conservation and packaging solutions, namely by the substitution of plastic polymeric material by edible formulations and by the substitution of synthetic additives by extracts of marine origin.

Carina Félix

Carina Félix (Ph.D.) is an assistant researcher in marine biotechnology working at Cetemares, Polytechnic of Leiria. She is currently studying bioactive compounds of invasive seaweeds and their possible applications in several fields, from agriculture to cosmetics. Holds a Ph.D. in biology by the University of Aveiro, with a specialization in microbiology, having extent experience in multi-omics functional analyses of phytopathogenic fungi.

Editorial

Marine Resources Application Potential for Biotechnological Purposes

Marco F. L. Lemos *[id], Sara C. Novais *[id], Susana F. J. Silva * and Carina Félix *

MARE–Marine and Environmental Sciences Centre, ESTM, Politécnico de Leiria, 2050-641 Peniche, Portugal
* Correspondence: marco.lemos@ipleiria.pt (M.F.L.L.); sara.novais@ipleiria.pt (S.C.N.);
   susana.j.silva@ipleiria.pt (S.F.J.S.); carina.r.felix@ipleiria.pt (C.F.)

Citation: Lemos, M.F.L.; Novais, S.C.; Silva, S.F.J.; Félix, C. Marine Resources Application Potential for Biotechnological Purposes. *Appl. Sci.* **2021**, *11*, 6074. https://doi.org/10.3390/app11136074

Received: 26 April 2021
Accepted: 28 June 2021
Published: 30 June 2021

Publisher's Note: MDPI stays neutral with regard to jurisdictional claims in published maps and institutional affiliations.

1. Marine Resources Application Potential for Biotechnological Purposes

Blue biotechnology plays a major role in converting marine biomass into societal value; therefore, it is a key pillar for many marine economy developmental frameworks and sustainability strategies, such as the Blue Growth Strategy, diverse Sea Basin Strategies (e.g., Atlantic Action Plan Priority 1 and 2 and COM (2017) 183), the Marine Strategy Framework Directive, the Limassol Declaration, or even the UN Sustainable Development 2030 Agenda. However, despite the recognized biotechnological potential of marine biomass, the work is dispersed between multiple areas of applied biotechnology, resulting in few concrete examples of product development. Food and feeds, and high-revenue cosmeceutical, pharma, biomedical markets, and others, are increasingly becoming important for marine bio compounds, which hold a myriad of unexploited uses, because they have often been demonstrated to contain molecules with a plethora of bioactivities, ranging from antioxidant, anti-inflammatory, tissue-specific protection, antimicrobial, anti-tumoral, antifouling, to texturizing, among many others. Market-driven and industrially orientated research, which increases the efficiency of the marine biodiscovery pipeline and ultimately delivers realistic and measurable benefits to society, is thus paramount for sustained blue growth and contribution towards the successful market penetration of targeted biomolecules or enriched extracts for new product development, and ultimately, contribute to a myriad of the UN's Sustainable Development Goals.

2. This Special Issue

The present Special Issue covers review articles and research papers addressing the biological activities of marine resources which may present high applicability and potential for industrial purposes. From submissions, eight high-quality manuscripts were selected covering the above-mentioned topics.

The importance of marine micro- and macroalgae, yeast, and even bacteriophages in the applied biotechnology field is evident from the well-split number of manuscripts presented here.

The potential of microalgae strains as sustainable alternative sources for commercial lutein, known to help maintain normal visual function, was critically reviewed by Sushanta Saha and co-workers [1], who also address appropriate cultivation strategies and market challenges. The authors argue that microalgae can, in fact, be a competitive and sustainable natural source, presenting higher growth rates and not requiring arable land and/or a growth season, compared to marigold flowers, the predominant natural source of lutein.

Bernardo Duarte and co-workers [2] focused their work on the premise that diatoms are potential added-value biorefineries producing unique pigments, triglycerides (TAGs), and long-chain polyunsaturated fatty acids (LC-PUFAs), with potential applications for feed, food, or even biofuel industries. This manuscript describes how pigments and fatty acid production in the diatom *Phaeodactylum tricornutum* can be modulated by spectral light control during their growth and contribute towards a better cultivation of these organisms for biotechnological purposes, and additionally discuss how diatoms may use such mechanisms to efficiently regulate light absorption and cell buoyancy.

Mari Ruiz-Domínguez and co-workers [3] argue that the microalgae *Isochrysis galbana* offers a broad range of bioactive compounds such as carotenoids, PUFA, and antioxidants, but most of current extraction solutions use toxic solvents, which causes environmental concerns. In their work, they optimized a method for the enhanced extraction of bioactive compounds from this species, using green technology of supercritical fluid extraction and applying a Box–Behnken design, presenting this methodology as a relevant and sustainable alternative to obtain important functional ingredients to be used in food and nutraceuticals.

To obtain a sustainable alternative to fish oil and to find natural alternatives for pigments to be used in animal nutrition, Natalie Pino-Maureira and co-workers [4] propose Rhodotorula strains in their work, which were isolated from seawater, for the biosynthesis of docosahexaenoic acid (DHA) and canthaxanthin using a low-cost source of carbon. The results presented are the foundation to use sustainable and low-cost growth media, in a circular economy framework, to potentially scale the production and provide novel sources of these essential nutrients.

Two studies were submitted proposing the seaweed *Grateloupia turuturu* as a potential source of bioactive compound. This seaweed is considered an invasive species in Europe, although it is known to contain a myriad of bioactive secondary metabolites with different potentialities. Invasive species have damaging effects on the environment and, by adding value to this biomass, the harvesting industry will contribute to control and even restore impacted environments, thus turning these threats into opportunities [5].

Rafael Félix and co-workers [6] addressed the development of green and cost-effective extraction protocols to optimally obtain UV-shielding and antioxidant compounds from *G. turuturu*, using response surface methodology, and evaluated the effect of ethanol concentration, liquid–solid ratio, pH, temperature, and time to enhance bioactive compound extraction. They argue that these protocols will allow for an optimized and sustainable use of this marine resource, contributing to the potential development of natural and eco-friendly cosmeceuticals from low-cost biomass.

Grateloupia turuturu is, in fact, presented as a macroalgae with a vast array of applications due to the diversity of compounds with relevant bioactivities. Carina Félix and co-workers [7] further addressed this seaweed potential by considering antioxidant, UV absorbance, anti-enzymatic (elastase and hyaluronidase), antimicrobial, and anti-inflammatory activities, as well as photoprotection potential, and its promising uses in the cosmeceutical field. Several proposed extracts presented relevant potential, namely, antimicrobial and anti-inflammatory activities, highlighting *G. turuturu*'s potential for the further development of natural formulations for skin protection.

Extraction, harvesting and processing, and cultivation conditions, are key performers in the success of using a resource for biotechnological purposes. In their work, Glacio Araujo and co-workers [8] addressed the seaweed *Calliblepharis jubata*—an edible red seaweed and a carrageenan primary producer yet not valued by the industry—harvested from the wild, and in semi-controlled and controlled aquaculture conditions. They characterized fatty acids, carbohydrates, and carrageenan content, and argued about the differences between the biological compounds of interest from different sources, identifying the advantages for human consumption and the food industry.

Lihua Xu and co-workers [9], based on the knowledge of some viruses acting as therapeutic agents to treat infectious diseases, studied the Vibrio phage Yong-XC31 as an agent to control *Vibrio mediterranei* and prevent disease in aquaculture, controlling *Pyropia* seaweed vibriosis, and eventually other affected organisms such as corals or scallops. Characteristics and the complete genome of this phage were analyzed, and new insights are presented for the application of giant phages as an important biotechnological product.

3. The Future of Marine Resources Biotechnology

This Special Issue's manuscripts highlight the vast potential that marine resources hold, from viruses to seaweeds, and a myriad of applications from antimicrobials and cosmetics to feed and food. However, despite the increasing body of research targeting marine resources and their potential application, a market-driven framework will definitely boost the blue economy and provide increasing solutions that may further rely on the large share of unknown organisms that are yet to be discovered in our oceans. It is thus paramount to invest in the biodiscovery process but also to address the target industries and define methodologies that are easily transported to industrial-friendly setups, where sustainability based on cost-effectiveness, green processes and a circular economy is indeed a cornerstone issue for the present and the future of a marine biobased economy.

Funding: With support from FCT/MCTES through national funds (UID/MAR/04292/2020), the European Union through EASME Blue Labs project AMALIA, Algae-to-MArket Lab IdeAs (EASME/EMFF/2016/1.2.1.4/03/SI2.750419), project VALORMAR (Mobilizing R&TD Programs, Portugal 2020) co-funded by COMPETE (POCI-01-0247-FEDER-024517), RD&T co-promotion project ORCHESTRA (No. 70155), and SAICTPAC/0019/2015—LISBOA-01-0145-FEDER-016405 Oncologia de Precisão: Terapias e Tecnologias Inovadoras (POINT4PAC) through the European Regional Development Fund.

Acknowledgments: The guest editors wish to thank all involved authors for their manuscripts and contribution, the reviewers for their cornerstone role in the publishing process, and the *Applied Sciences* editors and editorial staff.

Conflicts of Interest: The authors declare no conflict of interest.

References

1. Saha, S.K.; Ermis, H.; Murray, P. Marine Microalgae for Potential Lutein Production. *Appl. Sci.* **2020**, *10*, 6457. [CrossRef]
2. Duarte, B.; Feijão, E.; Goessling, J.W.; Caçador, I.; Matos, A.R. Pigment and Fatty Acid Production under Different Light Qualities in the Diatom *Phaeodactylum tricornutum*. *Appl. Sci.* **2021**, *11*, 2550. [CrossRef]
3. Ruiz-Domínguez, M.C.; Cerezal, P.; Salinas, F.; Medina, E.; Renato-Castro, G. Application of Box-Behnken Design and Desirability Function for Green Prospection of Bioactive Compounds from *Isochrysis galbana*. *Appl. Sci.* **2020**, *10*, 2789. [CrossRef]
4. Pino-Maureira, N.L.; González-Saldía, R.R.; Capdeville, A.; Srain, B. *Rhodotorula* Strains Isolated from Seawater That Can Biotransform Raw Glycerol into Docosahexaenoic Acid (DHA) and Carotenoids for Animal Nutrition. *Appl. Sci.* **2021**, *11*, 2824. [CrossRef]
5. Pinteus, S.; Lemos, M.F.L.; Alves, C.; Neugebauer, A.; Silva, J.; Thomas, O.P.; Gaspar, H.; Botana, L.M.; Pedrosa, R. Marine Invasive Macroalgae: Turning a real threat into a major opportunity—The biotechnological potential of *Sargassum muticum* and *Asparagopsis armata*. *Algal Res.* **2018**, *34*, 217–234. [CrossRef]
6. Félix, R.; Carmona, A.M.; Félix, C.; Novais, S.C.; Lemos, M.F.L. Industry-Friendly Hydroethanolic Extraction Protocols for *Grateloupia turuturu* UV-Shielding and Antioxidant Compounds. *Appl. Sci.* **2020**, *10*, 5304. [CrossRef]
7. Félix, C.; Félix, R.; Carmona, A.M.; Januário, A.P.; Dias, P.D.M.; Vicente, T.F.L.; Silva, J.; Alves, C.; Pedrosa, R.; Novais, S.C.; et al. Cosmeceutical Potential of *Grateloupia turuturu*: Using Low-Cost Extraction Methodologies to Obtain Added-Value Extracts. *Appl. Sci.* **2021**, *11*, 1650. [CrossRef]
8. Araujo, G.S.; Cotas, J.; Morais, T.; Leandro, A.; García-Poza, S.; Gonçalves, A.M.M.; Pereira, L. *Calliblepharis jubata* Cultivation Potential—A Comparative Study between Controlled and Semi-Controlled Aquaculture. *Appl. Sci.* **2020**, *10*, 7553. [CrossRef]
9. Xu, L.; Li, D.; Tong, Y.; Fang, J.; Yang, R.; Qin, W.; Lin, W.; Pan, L.; Liu, W. A Novel Singleton Giant Phage Yong-XC31 Lytic to the *Pyropia* Pathogen *Vibrio mediterranei*. *Appl. Sci.* **2021**, *11*, 1602. [CrossRef]

Review

Marine Microalgae for Potential Lutein Production

Sushanta Kumar Saha *, **Hande Ermis** and **Patrick Murray**

Shannon Applied Biotechnology Centre, Limerick Institute of Technology, Moylish Park, V94 E8YF Limerick, Ireland; Hande.Ermis@lit.ie (H.E.); Patrick.Murray@lit.ie (P.M.)
* Correspondence: Sushanta.Saha@lit.ie; Tel.: +353-61-293-536

Received: 8 September 2020; Accepted: 13 September 2020; Published: 16 September 2020

Abstract: Lutein is particularly known to help maintain normal visual function by absorbing and attenuating the blue light that strikes the retina in our eyes. The effect of overexposure to blue light on our eyes due to the excessive use of electronic devices is becoming an issue of modern society due to insufficient dietary lutein consumption through our normal diet. There has, therefore, been an increasing demand for lutein-containing dietary supplements and also in the food industry for lutein supplementation in bakery products, infant formulas, dairy products, carbonated drinks, energy drinks, and juice concentrates. Although synthetic carotenoid dominates the market, there is a need for environmentally sustainable carotenoids including lutein production pathways to match increasing consumer demand for natural alternatives. Currently, marigold flowers are the predominant natural source of lutein. Microalgae can be a competitive sustainable alternative, which have higher growth rates and do not require arable land and/or a growth season. Currently, there is no commercial production of lutein from microalgae, even though astaxanthin and β-carotene are commercially produced from specific microalgal strains. This review discusses the potential microalgae strains for commercial lutein production, appropriate cultivation strategies, and the challenges associated with realising a commercial market share.

Keywords: commercial microalgae cultivation; dietary supplements; lutein production; marine microalgae

---

1. Introduction

Microalgae are photosynthetic microscopic organisms that possess several accessory light-harvesting carotenoid pigment molecules such as astaxanthin, canthaxanthin, lutein, zeaxanthin, and β-carotene, which have commercial value. Lutein is a xanthophyll and one of 1178 known naturally occurring carotenoids [1]. It is an oxygenated carotenoid found primarily in plants such as spinach, kale, and marigold as well as certain microalgal species such as *Scenedesmus almeriensis*, *Chlorella zofingiensis*, and *Muriellopsis* sp. [2]. Lutein is a lipid-soluble primary carotenoid that humans obtain from their diet and has several known health benefits including aiding in the prevention of macular degenerative disease, reducing the risk of stroke and heart attack, and mitigation against other debilitating metabolic syndromes. In photosynthetic species, xanthophylls act to modulate light energy and free radical quenching agents which are produced during photosynthesis under high light intensity. Lutein is found to accumulate in the macula of the eye, acting as a light filter protecting cells against free radical damage, and has also been implicated in ameliorating the damaging effects of macular degenerative disease in ageing adults [3]. These health-promoting effects of lutein as well as its potential as a natural food colourant have led to increased investigations on the potential of lutein as a high-value nutraceutical functional food ingredient. The growth of the global nutraceutical market is driven by an increase in demand for healthy and organic food products and a surge in awareness of dietary health promoting supplements. Furthermore, the rise in disposable income allows consumers to purchase healthy alternatives to regular food products. The global lutein market is expected to reach at EUR 409 million by 2027 at a Compound Annual Growth Rate (CAGR) of 6.10% over the predicted period 2020–2027 [4].

Primarily, commercial natural lutein production has been reliant on extraction from marigold flower oleoresin. However, marigold flower harvesting and extraction is seasonal and labour intensive and recent data have suggested that microalgal species under controlled cultivation conditions can have much higher lutein productivity rates when compared to marigold cultivars [5]. There is, therefore, the strong potential for these organisms to be an alternative production route for natural lutein. Microalgae are attractive lutein producers; concomitantly, they function as a carbon dioxide capturing system reducing greenhouse gases, they can be cultivated all year round depending on the selected reactor cultivation conditions chosen, and environmentally friendly solvent-free extraction strategies can be tailored to enriched-lutein extraction e.g., supercritical fluid extraction methodologies [6].

The use of microalgae as human food is not unusual as it can be traced back many years in indigenous populations from China, Japan, and the Republic of Korea [7]. The traditional knowledge of microalgae use by these indigenous people has now disseminated throughout the world population through migration. Further, growth in microalgal consumption has been due to the significant amount of research on the health and nutritional benefits of microalgae and these health benefits are especially relevant for our modern-day lifestyle [8]. As a result, presently, several microalgae are commercially cultivated for various nutraceutical products that are available in the market. *Arthrospira* (*Spirulina*) *maxima*, *Arthrospira* (*Spirulina*) *platensis*, and *Aphanizomenon flos-aquae* are used as "whole cells biomass" in supplement products as well as for extraction of blue food colourant phycocyanin, *Chlorella vulgaris* biomass for health supplement products, and *Dunaliella salina* and *Haematococcus pluvialis* for commercial production of natural carotenoids such as β-carotene and astaxanthin [9].

2. Health Benefits of Lutein

Lutein, also known as the "eye vitamin", functions as a light protector by sheltering the eye tissues from harmful sunlight damage. The natural lens in the eye, which must remain clear, mainly collects and focuses light on the retina, where oxidation of this lens (clouding of the lens) is a major cause of cataracts. Antioxidants, such as lutein, neutralise free radicals that are connected with oxidative stress and retinal damage [10]. Because of its antioxidant potential, lutein is gaining popularity as a functional food ingredient to prevent age-related macular degeneration (AMD) [11]. Moreover, lutein has beneficial health effects other than eye health including anti-inflammatory, anti-atherogenic, antihypertensive, antidiabetic, antiulcer, and anticancer effects [12]. Lutein is one of the two main carotenoids found in human milk [13] and is very important for infant visual and cognitive development [14]. Lutein is one of the predominant carotenoids found in the new-born brain—about 66–77% of the total carotenoids [15]. These health-promoting effects of lutein (Figure 1) as well as its potential as a natural food colourant have led to increased investigations on the potential of lutein as a high-value nutraceutical functional food ingredient [16]. However, the health claim related to the cause and effect relationship between the consumption of lutein and maintenance of normal vision has not been approved by the European Food Safety Authority (EFSA) at the European level. From the scientific evidence, EFSA, however, agree that lutein consumption can increase macular pigment density in most of the healthy subjects (https://efsa.onlinelibrary.wiley.com/doi/pdf/10.2903/j.efsa.2012.2716); therefore, the challenge to prove the health benefits of lutein for the approval of health claims remains open.

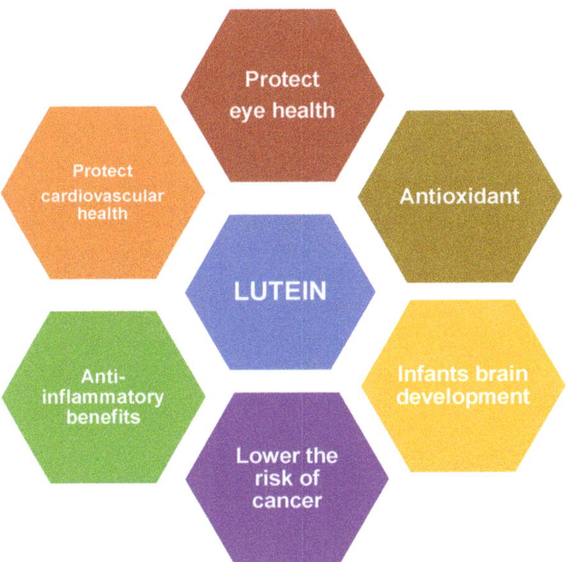

Figure 1. Schematic showing multiple potential health benefits of lutein.

3. Putative Biosynthetic Pathway of Lutein in Microalgae

Lutein is found within the members of Chlorophyta, Chlorarachniophyta, Cryptophyta, Euglenophyta, and Rhodophyta algal species [17]. Knowledge of the biosynthetic pathways for lutein biosynthesis in microalgae is limited. The current understanding is based on identified chemical structures of carotenoids found in microalgae. It is now believed that all types of carotenoids, including lutein, are obtained from common five-carbon (C_5) starting molecules isopentenyl diphosphate (IPP) and dimethylallyl diphosphate (DMAPP). These common metabolic precursors (IPP and DMAPP) might be derived from either of the two independent pathways: (i) the cytosolic mevalonate (MVA) pathway starting from Acetyl-CoA, or (ii) the plastidic (chloroplast in microalgae) methylerythritol 4-phosphate (MEP) pathway starting from pyruvate [18]. However, there is evidence to suggest that the precursors for microalgal carotenoids including lutein biosynthesis proceed from the MEP pathway in *Dunaliella salina, Chlorella vulgaris, Scenedesmus* sp. [18], and *Haematococcus pluvialis* [19] species.

Lutein biosynthesis in microalgae begins in the MEP pathway (Figure 2) through the 5-carbon building block molecule isopentenyl pyrophosphate (IPP), which isomerises to dimethylallyl pyrophosphate (DMAPP) by the action of the enzyme IPP isomerase. Elongation of the carbon chain then takes place through continuous head-to-tail condensation of IPP to DMAPP followed by growing of the polyprenyl pyrophosphate chain by the action of the enzyme prenyltransferase [20–22]. As a result, geranylgeranyl PP (GGPP, C_{20}), the first immediate precursor of lutein, is synthesised following a condensation reaction of GPP (C_{10}) by the action of GGPP synthase. Next, the colourless C_{40} carotenoid phytoene is formed through condensation of two GGPP (C_{20}) molecules by the action of phytoene synthase (PSY). Next, lycopene (a coloured carotenoid with 13 double bonds and a chromophore of 11 conjugated double bonds) is formed by the conversion of phytoene (nine double bonds molecule) through stepwise desaturation reactions (or dehydrogenation reactions) catalysed by phytoene desaturase (PDS) and zeta-carotene desaturase (ZDS) enzymes. Lycopene is the precursor to the formation of both α-carotene and β-carotene after two cyclisation reactions. The enzyme lycopene β-cyclase (*lcy-b*) catalyses the cyclisation of both ends of lycopene to make β-carotene with two β-rings. Meanwhile, the action of lycopene β-cyclase (*lcy-b*) and lycopene ε-cyclase (*lcy-e*) enzymes catalyse the cyclisation of both ends of lycopene to make α-carotene with a β-ring and an ε-ring. This is an important branching point of the carotenoid biosynthesis pathway in microalgae, where one branch leads to the biosynthesis of α-carotene and the other branch leads to the biosynthesis of β-carotene. The former is then converted to lutein in two hydroxylation steps, while the latter is converted to zeaxanthin and subsequently, to other carotenoids. The hydroxylation of α-carotene at the C-3 and C-3′ positions results in the formation of lutein and the enzymes involved in these processes are β-carotene hydroxylase and ε-carotene hydroxylase, respectively [23].

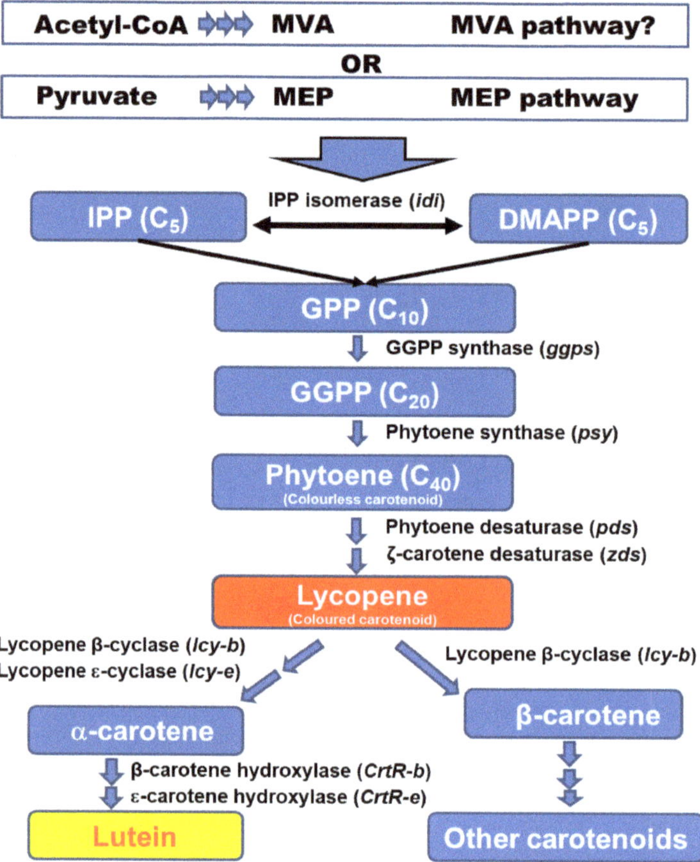

Figure 2. Schematic overview of lutein biosynthesis in microalgae. Enzymes involved in each biochemical conversion step are listed and their corresponding genes are indicated in parenthesis.

From the biotechnological perspective, a number of chemical inhibitors have been tested to regulate the carotenoid biosynthetic pathway in microalgae. For examples, Yildirim et al. [24] studied the effect of the addition of 2-methylimidazole in *Dunaliella salina* and found an increase of 1.7-fold lutein content and a related decline in β-carotene content. This study suggested that 2-methylimidazole preferentially alters lycopene β-cyclase (*lcy-b*) activity and thus, shifts the carotenogenic pathway from β-carotene to the α-carotene branch. Liang et al. [25] tested triethylamine, which triggered lycopene production in *Dunaliella bardawil* as a lycopene cyclase inhibitor that inhibited the expression levels of *lcy-b* and *lcy-e*, and upregulated the upstream carotenogenic genes. Likewise, nicotine was also tested as a possible lycopene cyclase inhibitor. A low concentration of nicotine resulted in a significant decrease in β-carotene, while triggering the accumulation of lycopene in *Chlorella regularis* Y-21 and *Dunaliella salina* CCAP 19/18 [26,27].

4. Engineering of Biosynthetic Pathways in Microalgae for Lutein Production

Microalgae as biomolecule production platforms have long been explored through engineering of their biosynthetic pathways by chemical mutagenesis as well as targeted genetic engineering of specific genes. Chemical mutagens such as N-methyl-N'-nitro-N-nitrosoguanidine (MNNG) and Ethyl Methane Sulfonate (EMS) have been successfully used for generating microalgal mutants with high contents of carotenoids including lutein [21,28–30]. In a recent study, a lutein-deficient *Chlorella vulgaris* (CvLD) strain was generated by chemical random mutagenesis and the strain was identified as an enhanced producer of violaxanthin [21]. Sequence analysis of the lycopene ε-cyclase gene (*lcy-e*) of the mutant strain CvLD revealed a single mutation (A336V), which might have resulted in a conformational change of the conserved region of CvLCYE (lycopene ε-cyclase) with reduced biochemical activity.

For genetic engineering of microalgae, robust genetic transformation protocols for the nuclear, chloroplast, or mitochondrial genomes are available [31]. The key methods for delivering DNA to microalgae are electroporation, shaking of cells with glass beads, and the biolistic (particle gun bombardment) method [32]. Most of the enzymes of secondary metabolism are encoded within the nuclear DNA; however, some of these enzymes are actually targeted to function in the chloroplast [33]. Since most of the enzymes of the carotenogenic pathway are found in the microalgal chloroplast, it was suggested that nuclear and/or chloroplast transformation can be used for the metabolic engineering of the carotenoid biosynthetic pathway [34]. Genetic engineering of microalgae through chloroplast transformation was mostly achieved by the biolistic method [35]. Although the low expression level of the target gene is the main drawback of nuclear transformation, it is the method of choice when appropriate post translational modifications of the target protein are essential [31].

The phytoene synthase gene (*psy*) from *Dunaliella salina* [28] and *Chlorella zofingiensis* [36] was nuclear transformed to the model microalga *Chlamydomonas reinhardtii*, and the derivative strains, respectively, produced 2.6- and 2.2-fold higher amounts of lutein compared to the wild type strain. Another study with the microalga *Chlamydomonas reinhardtii* involved a point mutation of the native gene for phytoene desaturase (*pds*) which was then nuclear transformed. The derivative strain, consisting of the mutant enzyme with increased desaturase activity biosynthesised, increased the amount of lutein, β-carotene, zeaxanthin, and violaxanthin [37]. In a very similar approach, Liu et al. [38] nuclear-transformed *Chlorella zofingiensis* with a mutant version of the native *pds* gene. The derivative strain, consisting of the mutant phytoene desaturase enzyme with increased desaturase activity, accumulated a higher amount of total carotenoids (32%) and astaxanthin (54%) compared to the parent strain. In a recent metabolic engineering study with *Chlamydomonas reinhardtii*, the gene for phytoene-β-carotene synthase (*pbs*) from the red yeast *Xanthophyllomyces dendrorhous* was cloned into pMS188 vector and nuclear transformed [39]. This derivative strain possesses the bifunctional enzyme with both phytoene synthase (*psy*) and lycopene cyclisation (*lcy-b*) activities. This is the first heterologous expression system for carotenoids biosynthesis, which resulted in a simultaneous increase in lutein (60%) and β-carotene (38%) under low light conditions.

In the lutein biosynthetic pathway, the lycopene ε-cyclase gene (*lcy-e*) acts as a key regulator of the α-branch, while lycopene β-cyclase (*lcy-b*) acts in both the α-branch and β-branch. The findings from chemical mutagenesis and genetic engineering studies suggest that an altered lycopene β-cyclase with enhanced activity in α-branch and reduced or no activity for β-branch, as well as an altered lycopene ε-cyclase with enhanced activity, might be a target for creating a genetically engineered strain with enhanced lutein productivity. Additionally, the cytosolic MVA pathway of microalgae can be genetically engineered to express a whole heterologous metabolic pathway for lutein biosynthesis. This approach was recently successfully demonstrated in tobacco (*Nicotiana tabacum* L.) plant, where a viral vector was used to express three enzymes (GGPP synthase (*crtE*), phytoene synthase (*crtB/psy*), and phytoene desaturase (*crtI*)) from the soil bacteria *Pantoea ananatis* to biosynthesise lycopene [40].

5. Synthetic Production of Lutein

Since the development of the chemical synthesis method for carotenoids production in 1950, β-carotene has long been chemically synthesised industrially for meeting global demand [41]. The chemically synthesised carotenoids list has been growing and includes astaxanthin, canthaxanthin, lycopene, zeaxanthin, and β-carotene [42,43]. The commercial production of lutein through the chemical synthesis method has not been viable due to poor overall yield. However, the technology news of the University of Maryland, USA reported that their researchers have published a chemical synthesis method for lutein with overall yields >20% from readily available precursor molecules [43,44]. At present, most of the commercially produced carotenoids are produced chemically and only a small portion are obtained from the biotechnological process, including extraction from microalgae, but not lutein by either method. Considering the fact that there is high demand and consumer preference for natural carotenoids, there is a huge opportunity for natural lutein production from microalgae.

6. Microalgae Cultivation for Commercial Lutein Production and Challenges

There are several microalgae species both from marine and freshwater habitats identified as potential lutein producers. Some of these species could produce up to 5 g of free lutein per kg biomass [16]. Microalgae are considered as important industrial candidates for lutein bioproduction. They have higher biomass productivity, high lutein content, are suitable for cultivation in freshwater, wastewater, brackish, or seawater, and/or are dependent neither on arable lands nor on local weather. However, the main challenge in carotenoid production from microalgae in general is low biomass growth rate due to the stress conditions necessary for carotenogenesis. Therefore, large-scale microalgal cultivation mostly operates in two steps, which are the cultivation of microalgae in optimum conditions for fast growth in complete growth medium followed by stress conditions to improve the desired carotenoid in microalgae [45]. There are many strategies to improve microalgal lutein content such as different light intensities, light colour, growth conditions, and nutrient limitations. In the study by Ho et al. [46], *Scenedesmus obliquus* FSP-3 was subjected to light-related strategies to increase cell growth and lutein production, where the best lutein productivity (4.08 mg L^{-1} d^{-1}) was observed at a light intensity of 300 μmol photons m^{-2} s^{-1} with white light. In another study by Ho et al. [47], the effects of nitrogen sources on the cell growth and lutein content of *Scenedesmus obliquus* FSP-3 were examined, where the highest lutein content (4.61 mg g^{-1}) and lutein productivity (4.35 mg L^{-1} d^{-1}) were obtained when using 8.0 mM calcium nitrate as the nitrogen source. After determining the nitrogen source condition, Ho et al. [47] applied two bioreactor strategies which were semi-continuous and two-stage; and observed that semi-continuous operation with a 10% medium replacement ratio achieved the highest biomass productivity (1304.8 mg L^{-1} d^{-1}) and lutein productivity (6.01 mg L^{-1} d^{-1}).

Zhao et al. [48] examined the marine microalga *Chlamydomonas* sp. JSC4 under different environmental conditions for lutein production and concluded that the optimal lutein content was obtained under blue light and a lower temperature of 20–25 °C. Schüler et al. [49] also examined different environmental factors on marine microalga *Tetraselmis* sp. CTP4 for carotenoid biosynthesis and examined that lutein amount increased 1.5-fold under higher light intensities (170 and 280 μmol photons m^{-2} s^{-1}), but in contrast to Zhao et al. [48], found a temperature increase from 20 to 35 °C, after only two days at a light intensity of 170 μmol photons m^{-2} s^{-1}, yielded the highest lutein productivity (3.17 ± 0.18 mg g^{-1} dry weight biomass). Ma et al. [50] also examined light stress on the marine microalga *Chlamydomonas* sp. JSC4 as a potential lutein production. High lutein productivity (5.08 mg $L^{-1}d^{-1}$) was attained under high light irradiation of 625 μmol photons m^{-2} s^{-1}, where lutein amount started to decrease at 750 μmol photons m^{-2} s^{-1} due to downregulation of *lut1* and *zep* genes, which respectively encode the enzymes ε-carotene hydroxylase and zeaxanthin epoxidase that are responsible for lutein biosynthesis.

Chen et al. [51] also examined light-related carotenogenesis strategies along with different nitrogen concentrations and growth conditions on *Scenedesmus obliquus* CWL-1 under mixotrophic cultivation. In contrast to Ho et al. [46], maximum lutein yield (1.43 mg L^{-1} d^{-1}) was achieved in 12 h/12 h light/dark conditions under blue/red light. In addition to the light strategy, the addition of 4.5 g L^{-1} of calcium nitrate increased lutein productivity (3.06 mg L^{-1} d^{-1}), while the addition of 1.5 g L^{-1} of calcium nitrate increased the lutein content to 2.45 mg g^{-1}. Finally, compared to a batch cultivation system, a fed-batch cultivation strategy had 11-fold higher lutein productivity (4.96 mg L^{-1} d^{-1}), which was the highest productivity that was achieved compared to all strategies.

In the study by Florez-Miranda et al. [52], a two-stage cultivation strategy was applied, where heterotrophy with different temperatures was followed by photoinduction to improve biomass and lutein production in *Scenedesmus incrassatulus*. In the initial stage, where different nitrogen sources and temperatures were applied, the highest lutein content was observed using urea plus vitamins at 30 °C; during the second stage, after 24 h of photoinduction, lutein content increased seven times. According to the comparison between autotrophic and heterotrophic growth, Florez-Miranda et al. [52] concluded that lutein productivity was 1.6 times higher compared to autotrophic cultivation. Hence, heterotrophic and/or mixotrophic growth for lutein production can be suggested as a better cultivation method compared to autotrophic growth. Xie et al. [53] observed different types of media and concentrations of sodium acetate and nitrate on *Chlorella sorokiniana* FZU60 to improve mixotrophic growth and lutein production, where highest lutein content (9.57 mg g^{-1}) and productivity (11.57 mg L^{-1} d^{-1}) were obtained in BG-11 medium supplemented with 1 g L^{-1} acetate and 0.75 g L^{-1} nitrate. Moreover, pulse feeding with 1 g L^{-1} acetate every 48 h led to the alternation between mixotrophy and photoinduction, which resulted in a lutein production of 33.6 mg L^{-1}.

In the study by Shi et al. [54], *Chlorella protothecoides* was cultivated heterotrophically (40 g L^{-1} glucose and 3.6 g L^{-1} urea) with nitrogen limitation in a fed-batch culture, which was followed by an increased lutein production without drastically reducing the biomass amount. Afterwards, this N-limited fed-batch culture was successfully scaled up from 3.7 to 30 L, and temperature stress was applied. Higher temperature (32 °C) for 84 h resulted in a 19.9% increase in lutein content but a 13.6% decrease in biomass amount, as compared to the fed-batch culture (30 L) without any treatment.

In the study by Shinde et al. [55], the lutein content of *Auxenochlorella protothecoides* SAG 211-7a was observed under heterotrophic conditions, where the Taguchi Orthogonal Array method, a statistical technique for screening and optimisation of medium components at small-scale, was applied to select the six independent variables (yeast extract, KH$_2$PO$_4$, MgSO$_4$.7H$_2$O, CaCl$_2$.2H$_2$O, EDTA, and pH) that affect lutein production. Sucrose, yeast extract, MgSO$_4$.7H$_2$O, and EDTA were selected as the significant factors affecting lutein production. According to the experimental results, 1303 ± 25.32 µg L^{-1} lutein was produced when the medium was supplemented with 14 g L^{-1} sucrose, 3 g L^{-1} yeast extract, 0.8 g L^{-1} MgSO$_4$.7H$_2$O, and 0.76 g L^{-1} EDTA.

In conclusion, compared to all strategies mentioned above, the best approach for a microalgae cultivation system for maximum lutein productivity is a fed-batch cultivation strategy with different light and temperature stress depending on the selected freshwater culture and salinity for marine culture, since the maximum level of lutein in the cell is dependent on the salinity [56].

There is no commercial production system for lutein from microalgae yet; however, some outdoor productions of *Muriellopsis* sp. and *Scenedesmus* sp. at a pilot scale have been created, where *Muriellopsis* sp. was cultivated in a 55 L tubular photobioreactor and the highest lutein productivity was 40 g m^{-2} d^{-1}. Moreover, *Scenedesmus almeriensis* was also cultivated in a 4000 L tubular photobioreactor for lutein production and 290 mg lutein m^{-2} d^{-1} was observed [57]. Besides lutein, the main commercial carotenoid productions are β-carotene from *Dunaliella* sp. and astaxanthin from *Haematococcus* sp. Commercially, β-carotene from *Dunaliella* sp was extracted by Betatene (Australia) and Western Biotechnology (Australia) in 400 and 240 ha extensive unmixed ponds, where 13–14 and 4–5 tonnes of β-carotene per year have been extracted, respectively. For commercial astaxanthin extraction, Cyanotech (Hawaii) uses a closed tubular photobioreactor followed by open ponds, whereas Algatechnologies (Israel) applies a closed tubular photobioreactor for *Haematococcus* sp. cultivation [57]. In conclusion, a tubular photobioreactor can be suggested for commercial lutein production as was experimentally tested for *Muriellopsis* and *Scenedesmus* species, and also due to the fact that tubular photobioreactors have been successfully used for other carotenoid-producing microalgae.

7. Extraction of Lutein from Microalgae and Challenges

The extraction of intracellular microalgae products is most commonly conducted by conventional means. This involves using dry biomass coupled with maceration and thermal extraction employing either organic or aqueous solvents (Table 1), depending on the polarity of the desired compound to be extracted. Compounds, such as carotenoids, display varying polarities and chemical natures. Hence, an appropriate solvent must be chosen with regards to extracting target carotenoids based on the selectivity, efficiency, and purity required. The extraction of carotenoids is typically conducted using non-polar solvents such as n-hexane, dichloromethane, and dimethyl ether, with other solvents such as acetone and octane also used for extraction [58].

The use of these non-polar solvents is due to the high hydrophobicity of carotenoids. Recently, green solvents such as ethanol and biphasic water solvent mixtures have also been researched for the extraction of carotenoids from microalgae [59]. The conventional extraction process of microalgae products still bears limitations such as extraction efficiency, selectivity, and high solvent consumption. Lin et al. [5] discussed that the energy consumption for microalgal cell disruption ranged from 30 to 500 MJ kg^{-1}. This energy consumption was reliant on the disruption process, where the energy demand is determined by factors such as cell wall compositions, cell wall thickness, and cell size. This value is 1000 times higher than the crushing energy for marigold flower (800 kJ kg^{-1}). To overcome this problem, the development and application of a multistage extraction process, combined with varying physical and chemical methods, may prove useful in selectively targeting the carotenoids of interest [58]. Microalgal biomass consists of a rich cellular composition and much thicker cell wall than marigold flowers. Dry-milled marigold petals are usually processed using solvent extraction to produce oleoresin-containing carotenoids in their ester form. This is followed by a multistep purification process which frees the hydroxylated carotenoids from the accompanying fatty acids and finally, a recrystallisation process occurs which results in pure lutein/zeaxanthin. Lutein is sold in oily extracts ranging from 5% to 60% as its crystalline form poses management and stability problems, where microalgal biomass can be extracted into an oleoresin-like extract with 25% lutein in free form that could be used directly for the commercial products [60].

Araya et al. [61] applied two cell disruption methods, glass bead vortexing and ball mill grinding, on *Chlorella vulgaris*, *Chlorella zofingiensis*, and *Chlorella protothecoides* to improve the extraction yield of lutein, where the yield of *C. vulgaris* (0.51 mg L^{-1} d^{-1}) and *C. zofingiensis* (0.53 mg L^{-1} d^{-1}) was higher compared to *C. protothecoides* (0.37 mg L^{-1} d^{-1}). Chen et al. [62] also applied two different cell disruption methods (bead-beating and high pressure) on *Chlorella sorokiniana* MB-1 and the lutein was extracted by a reduced pressure extraction method. High-pressure pretreatment extracted with tetrahydrofuran (THF) as the solvent resulted in high lutein recovery efficiencies of 87.0% (at 20 min incubation) and 99.5% (at 40 min incubation) at 850 mbar pressure and at temperature 25 °C.

Solvents such as hexane and/or ethanol are the most commonly employed methods for lutein extraction due to the easy removal of the solvent from the extract while also retrieving a high content of lutein. In contrast, direct extraction with vegetable oil was described by Nonomura [63] and does not allow for solvent removal. This patent involves direct extraction on wet biomass using the addition of vegetable oil, followed by emulsification and a resting period. However, in regard to microalgae with thick cell walls, such as *Murielopsis* sp. or *Scenedesmus* sp., it is unlikely this methodology would be efficient.

Low et al. [64] described a microwave-assisted binary phase solvent extraction method (MABS) for the recovery of lutein from microalgae. The method was established and optimised to specifically attain the highest lutein recovery from microalgae *Scenedesmus* sp. biomass, with a total of 11.92 mg g^{-1} lutein recovered. The optimal binary phase solvent composition was a 60% potassium hydroxide solution with acetone in the ratio of 0.1 (mL/mL). The highest lutein content was at 55 °C treatment temperature,

36 min extraction time, 0.7 (mg mL^{-1}) biomass:solvent ratio, 250 Watt microwave power, and 250 rpm stirring speed. This optimised novel procedure can increase lutein recovery by approximately 130%, along with also shortening the overall extraction time by 3-fold.

According to an optimised extraction method of lutein from microalgal biomass conducted by Ceron et al. [65], it was concluded that cell disruption was necessary and the use of a bead mill with alumina in a 1:1 (*w/w*) proportion as a disintegrating agent for 5 min was the preferred method amongst other varying treatments tested.

Studies conducted by Gong et al. [66] investigated lutein extraction from wet *Chlorella vulgaris* UTEX 265, where various extraction parameters such as sample size, drying method, and cell disruption method were studied and lutein production was monitored throughout the microalgal growth phase. From the analysis of varying solvent performances on lutein extraction using Nile Red as a solvatochromic polarity probe, it was determined that 3:1 (*v/v*) ethanol/hexane was the optimal solvent for lutein extraction, which resulted in 13.03 mg g^{-1} lutein.

Studies carried out by Li et al. [67] involved the use of high-speed counter-current chromatography (HSCCC) for the isolation and purification of lutein from microalgae. Analytical HSCCC was used for the preliminary selection of a suitable solvent system composed of n-hexane:ethanol:water (4:3:1, *v/v*), where HSCCC was successfully performed, resulting in lutein yield at 98% purity from 200 mg of crude extract in a one-step separation.

The use of supercritical CO_2 fluid is gaining increasing interest in regard to the recovery of pharmaceutical or nutraceutical compounds due to clean extracts and lack of toxicity of CO_2 in comparison to organic solvents. However, because the polarity of free lutein is high and excess amounts of organic solvents are required for lutein extraction from microalgae due to binding of light-harvesting complexes, supercritical fluid extraction (SFE) should not prove efficient. However, according to Yen et al. [68], the addition of polar co-solvents such as methanol or ethanol may increase lutein extraction efficiency. When coupled with a co-solvent, SFE may seem a promising lutein extraction method compared to organic solvent extraction. Miguel et al. [69] recommended the use of supercritical CO_2 following a solvent extraction of the carotenoids. The solvent containing the carotenoids was then mixed with supercritical CO_2 and the conditions of pressure and temperature were adjusted to promote the precipitation of lutein.

Macías-Sánchez et al. [60] aimed to clarify the influence of temperature and pressure parameters on the supercritical fluid extraction of lutein and β-carotene from a freeze-dried powder of the marine microalga *Scenedesmus almeriensis*. The extracts were analysed by HPLC and empirical correlations were also established. The results determined an optimal pressure of 400 bar and optimal temperature of 60 °C resulted in a significant yield of pigment extraction. In comparison with the referenced extraction processes used, the results obtained from this study displayed that improved yields were obtained in the extraction of β-carotene, where it was possible to extract 50% of the total of this pigment contained in the microalga studied.

Di Sanzo et al. [70] also investigated the use of supercritical fluid extraction using CO_2 for the extraction of astaxanthin and lutein from disrupted red phase biomass of the *Haematococcus pluvialis*. A bench-scale reactor in a semi-batch configuration was employed. Parameters such as extraction time (20, 40, 60, 80, and 120 min), CO_2 flow rate (3.62 and 14.48 g min^{-1}) temperature (50, 65, and 80 °C), and pressure (100, 400, and 550 bar.) were investigated. The results indicate the maximum recovery of astaxanthin and lutein obtained was 98.6% and 52.3%, respectively, at 50 °C and 550 bars.

Mehariya et al. [71] employed supercritical fluid for the extraction of lutein from *Scenedesmus almeriensis*. The optimisation of the main parameters affecting the extraction, such as biomass pretreatment, temperature, pressure, and CO_2 flow rate, was conducted. Firstly, the effect of mechanical pretreatment with diatomaceous earth (DE) and biomass mixing in the range of 0.25–1 DE/biomass, grinding speed varying between 0–600 rpm, and pretreatment time changing from 2.5 to 10 min, was evaluated on lutein extraction efficiency. Next, the influence of SFE extraction parameters such as pressure (250–550 bar), temperature (50 and 65 °C), and CO_2 flow rate (7.24 and 14.48 g min^{-1}) on lutein recovery and purity was examined. According to the results, increases in the temperature, pressure, and CO_2 flow rate improved the lutein recovery and purity. The maximum lutein recovery (~98%) with a purity of ~34% was accomplished running at 65 °C and 550 bar with CO_2 flow rate of 14.48 g min^{-1}.

Research conducted by Yen et al. [68] in regard to utilising SFE for the extraction of lutein from *Scenedesmus* sp. biomass found that an increase in both pressure and temperature resulted in an increase in lutein yield, but an increase in temperature resulted in increased impurity within the HPLC profile. This increase in yield was not as significant when compared to the yields from conventional extraction methods. Optimum parameters for lutein recovery yield were determined as 400 bar pressure and temperature at 70 °C while using ethanol as the co-solvent at a ratio of 30 mol%. These parameters resulted in lutein recovery of 76.7%, which was compared to conventional methanol extraction methods.

Research carried out by Ruen-ngam et al. [72] involved the use of a pretreatment process using alcohol aimed at removing chlorophyll *a*, *b*, and β-carotene from *Chlorella vulgaris*. This process was developed to enhance the yield and selectivity of lutein in the extract obtained by subsequent supercritical fluid extraction. SFE was carried out in the pressure range 200 to 400 bar and the temperature range of 40 to 80 °C, with methanol and ethanol trialled as the co-solvents, with ethanol the most suitable for lutein extraction. Lutein yield within the extract increased with pressure but decreased with temperature. The optimal parameters for lutein recovery yield were determined as 400 bar and 40 °C, using ethanol as the co-solvent. Under these conditions, the maximum recovery percentage and selectivity percentage of lutein resulted in approximately 52.9 ± 0.02% and 43.1 ± 0.02%, respectively.

Fan et al. [23] aimed to extract lutein from *Chlorella pyrenoidosa* using ultrasound-enhanced subcritical CO_2 extraction (USCCE). As part of this research, parameters such as pretreatment process, pressure, temperature, CO_2 flow rate, and ultrasonic power were examined, alongside orthogonal analysis, to study the effects of varying parameters on lutein extraction yields. The developed USCCE method was conducted as part of this study and it was compared to other common extraction methods. Optimal extraction conditions were determined as temperature at 27 °C, pressure at 210 bar, 1.5 mL g^{-1} ethanol, and ultrasound power at 1000 W. Under these conditions, a maximum extraction yield of 1.24 mg lutein g^{-1} crude material was obtained and when compared to other methods, it was found that USCCE could result in increasing lutein extraction yield significantly at much lower extraction pressure and temperature.

Wu et al. [73] also applied supercritical fluid extraction followed by HPLC and LCMS analysis to lutein extraction from *Chlorella pyrenoidosa*. Under optimum conditions, temperature at 50 °C, pressure at 250 bar, and 50% ethanol as the co-solvent, the extraction yield recovery of lutein was 87.0%. From the results, it was concluded that SFE yielded in high purity lutein recovery and the process developed during this study might be suitable for the commercial production of lutein.

In conclusion, in regard to the large-scale production of lutein, only solvent extraction has, to date, achieved high degrees of efficiency and purity. However, new advances in methods and techniques such as selective precipitation with supercritical CO_2 and new advantageous solvents, such as ethyl lactate, which have been proposed for the extraction of other plant matter [74], may also be applied to microalgae and prove beneficial.

Table 1. List of lutein-producing microalgae and their lutein yield associated with tested cultivation and stress conditions.

Microalgae	Biomass	Cultivation Conditions	Lutein Yield	Stress Conditions	Extraction Methodologies	References
Marine cultures						
Chlamydomonas sp. JSC4	1271 mg L^{-1} d^{-1}	1-L glass photobioreactor	3.27 mg L^{-1} d^{-1}	Temperature (35 °C)	Solvent extraction	[75]
Chlamydomonas sp.	1500 mg L^{-1} d^{-1}	1-L glass photobioreactor	5.08 mg L^{-1} d^{-1}	Light intensity (625 µmol photons m^{-2} s^{-1})	Solvent extraction	[50]
Chlamydomonas sp. JSC4	560 mg L^{-1} d^{-1}	1-L glass photobioreactor	3.42 mg g^{-1}	Salinity gradient	Solvent extraction	[76]
Chlamydomonas sp. JSC4	490 mg L^{-1} d^{-1}	1-L glass photobioreactor	2.95 mg g^{-1}	Light wavelengths (blue light)	Solvent extraction	[50]
Chlamydomonas acidophila	-	Batch growth	20 mg L^{-1}	UV-A radiation (10 µmol photons m^{-2} s^{-1}), or heated at 40 °C.	Solvent extraction	[77]
Chlorella salina	-	3-L glass flask	2.92 mg g^{-1}	-	Microextraction coupled with ultrasonication	[78]
Dunaliella salina	2.2 g m^{-2} d^{-1}	Tubular photobioreactor	15.4 mg m^{-2} d^{-1}	None	Solvent extraction	[79]
Muriellopsis sp.	40 g m^{-2} d^{-1}	Outdoor tubular photobioreactor	6 mg g^{-1}	None	Solvent extraction	[80]
Muriellopsis sp.	12.9 g m^{-2} d^{-1}	Open ponds	100 mg m^{-2} d^{-1}	None	Solvent extraction	[57]
Tetraselmis sp. CTP4	-	5-L reactors	3.17 mg g^{-1}	Light intensity (170 and 280 µmol photons m^{-2} s^{-1}) and temperature (35 °C)	Solvent extraction	[49]
Freshwater cultures						
Chlorella minutissima	0.117 g L^{-1} d^{-1}	2-L airlift photobioreactor	5.58 mg g^{-1}	None	Solvent extraction	[81]
C. vulgaris, C. zofingiensis and *C. protothecoides*	0.131, 0.122, 0.103 g L^{-1} d^{-1} (respectively)	In indoor vertical alveolar panel photobioreactor	3.86, 4.38 and 3.59 mg g^{-1} (respectively)	None	Glass bead vortexing and ball mill grinding	[61]
Chlorella protothecoides	31.2 g L^{-1}	Heterotrophic growth in a 3.7-L fermenter	1.90 mg g^{-1}	80 g L^{-1} glucose addition	Solvent extraction	[82]

Table 1. *Cont.*

Microalgae	Biomass	Cultivation Conditions	Lutein Yield	Stress Conditions	Extraction Methodologies	References
			Freshwater cultures			
Chlorella pyrenoidosa	-	-	1.24 mg g^{-1}	None	Ultrasound-enhanced subcritical CO$_2$ extraction	[23]
Chlorella sorokiniana	1.98 g L^{-1} d^{-1}	Two-stage mixotrophic cultivation	7.62 mg L^{-1} d^{-1}	None	Solvent extraction	[83]
Chlorella vulgaris	-	Batch	3.36 mg g^{-1}	None	Ultrasound extraction with enzymatic pretreatment	[84]
Chlorella sorokiniana	2.4 g L^{-1}	Semi-batch mixotrophic cultivation.	5.21 mg g^{-1}	None	Reduced pressure extraction method.	[85]
Chlorella protothecoides	28.4 g L^{-1}	Heterotrophic batch growth in a 3.7-L fermenter	0.27 mg g^{-1}	Nitrogen limitation and high temperature	Mechanical method	[54]
Chlorella zofingiensis	7 g L^{-1}	Batch growth	4 mg g^{-1}	None	Solvent extraction	[86]
Desmodesmus sp.	939 mg L^{-1} d^{-1}	1-L glass vessel	5.22 mg L^{-1} d^{-1}	Different C/N ratios (1:1 and 150 mg L^{-1})	Solvent extraction	[87]
Muriellopsis sp.	5.37 g L^{-1}	Batch growth	29.8 mg L^{-1}	None	Solvent extraction	[88]
Scenedesmus incrassatulus	17.98 g L^{-1}	Two-stage heterotrophy photoinduction culture	1.49 mg g^{-1}	Glucose concentration increase (30.3 g L^{-1})	Solvent extraction	[52]
Scenedesmus sp. CCNM 1028	0.47 g L^{-1}	Batch growth (1L)	2.12 mg g^{-1}	Two-stage nitrogen starvation	Solvent extraction	[89]
Scenedesmus obliquus CWL-1	9.88 g L^{-1}	Mixotrophic cultivation	1.78 mg g^{-1}	Light-related strategies (12/12 L/D, blue to red light)	Solvent extraction	[51]
Scenedesmus almeriensis	0.95 g L^{-1}	Vertical bubble column photo-bioreactor	8.54 mg g^{-1}	Different CO$_2$ Content (3.0% *v/v*)	Accelerated solvent extraction	[90]
Scenedesmus sp.	1.1 g L^{-1}	20 L photobioreactor	1.794 mg g^{-1}	Different pressure and temperature in the SFE operation (400 bar, 70 °C and ethanol as the co-solvent)	Supercritical CO$_2$ extraction	[68]
Scenedesmus almeriensis	0.63 g L^{-1}	Bubble column photobioreactors (2.0 L)	3.6 mg L^{-1}	Salinity (5 g L^{-1})	Solvent extraction	[91]
Scenedesmus obliquus	2.44 g L^{-1}	1-L glass vessel	3.63 mg g^{-1}	Light-related strategies	Solvent extraction	[46]

8. Current Market Demand, Value and Sources

Currently, the global lutein market is valued at EUR 255 million and is expected to reach EUR 409 million by 2027 [4]. Lutein appears as a slightly lower value carotenoid compared to the market price for astaxanthin. As per ICIS market research, the price for 100% pure lutein may range from EUR 1688.00 to EUR 2532.00 per kg, while the available products with dry forms of lutein may range from EUR 126.00 to EUR 253.00 per kg, and the products with the liquid forms of lutein may range from EUR 422.00 to EUR 590.00 per kg (https://www.icis.com/explore/resources/news/2003/05/16/195956/lutein-eyes-robust-growth-in-food-and-nutraceuticals/ accessed on 12 September 2020). The growth of the global lutein market is driven by an increase in demand for healthy and organic food products and a surge in awareness towards dietary supplements. Furthermore, the rise in disposable income allows consumers to purchase healthy alternatives to regular food products.

Asia Pacific is projected to account for 23.2% of the global market. Developing nations such as China and India are expected to observe high growth. High occurrence of eye disorders coupled with the growing demand for dietary supplements is expected to accelerate the need [92]. Dietary supplements held 29% of the overall market share in 2019 and are expected to keep their dominance over the forecast period. Europe held 36.2% of the industry in 2018 and is projected to grow significantly in the coming years [92].

9. Overall Discussion and Future Prospects

Although marigold meets the global demand for lutein to some extent, there is still a huge opportunity to contribute to the global demand for natural lutein. This is where microalgae can play a significant role because there are several microalgae that produce 0.5–1.2% lutein of their cell dry weight [5]. Moreover, microalgae have more free lutein than marigold, which is preferable since it is easily absorbed compared to the esterified forms found in marigold flowers. In addition, there are several microalgal commercial technologies for cultivation, and optimum extraction of carotenoids is available. Importantly, microalgae are considered as one of the most promising biofactories, with 5–10 times higher growth rate than land plants, high-potential CO_2 scavenging, and their commercial cultivation technologies can be based on all types of water sources such as freshwater, brackish water, and seawater, and do not require arable land. Microalgae as commercial lutein producers are still waiting for the involvement of enthusiast entrepreneur biotechnologists who would be willing to adopt existing microalgal cultivation technologies. It is good that there are already a number of marine as well as freshwater microalgal strains identified as lutein producers through various research studies (Table 1), which can serve as a starting point.

The commercially optimum method and the safety of chemically synthesised lutein for human consumption are still questionable; therefore, the natural lutein market is gaining interest. Even though there is no commercial production of lutein, microalgae are an attractive source for the mass production of lutein due to their high growth rates and high pigment content. However; there are some challenges to be focused on that are mostly related to the cost [93]. Particularly, these are due to the current market price for lutein as well as the lutein yield in known microalgae being lower compared to the market price and yield of astaxanthin, which is currently produced economically and reliably from *Haematococcus pluvialis*. Although heterotrophic cultivation improves cell growth and lutein content, the cost of glucose and other carbon sources prevent microalgal lutein production from being commercially profitable [94]. Not only the carbon addition cost, but also the downstream processes such as harvesting and drying increase the cost, where energy saving approaches are essential for both processes. Moreover, to improve the growth capacity of the selected culture, high efficiency photobioreactors should be designed, where optimal growth conditions such as temperature or light can be applied. Based on the current cultivation technologies for *Dunaliella* sp. and *Haematococcus* sp. that were also found suitable at experimental-scale for *Muriellopsis* sp. and *Scenedesmus* sp., it appears that tubular photobioreactors can also be the choice for lutein production from these microalgae.

In addition, advanced metabolic engineering approaches may be applied to improve the lutein synthesis pathways in microalgae and increase its cellular accumulation to be commercially competitive. In summary, microalgal lutein has a great potential to be commercially produced with some challenges mentioned above to be overcome. In the future, better engineering reactor designs should be created for higher culture growth, new innovations should be applied for low-cost harvesting/drying processes, and selected cultures should be metabolically engineered (by chemical mutagenesis or targeted genetic engineering) to produce higher lutein without decreasing the culture biomass yield.

Overall, considering the growing demand for natural lutein, available potential microalgal strains, and their cultivation technologies available, it is high time to initiate industrial involvement along with some research and development activities for microalgal lutein production, along with other value-added bioproducts, using a biorefinery approach. The approach would at least need to achieve, through the rapid cultivation of selected microalgal strains by tweaking their abiotic growth factors to enhance lutein content followed by simple harvesting of biomass, extraction of lutein at affordable costs and valorisation of lutein-extracted biomass for additional bioproducts development.

Author Contributions: Conceptualisation, S.K.S. and P.M.; methodology, S.K.S. and H.E.; formal analysis, S.K.S. and H.E.; investigation, S.K.S. and H.E.; writing—original draft preparation, S.K.S., H.E., and P.M.; writing—review and editing, S.K.S. and P.M. All authors have read and agreed to the published version of the manuscript.

Funding: Recent literature search and analysis were carried out in conjunction with one of the ongoing funded project (19/KGS/006) on the "Development of commercialisation pipeline of Microalgal bioFACTORIES starting from biodiscovery screening (M-FACTORIES)".

Acknowledgments: All authors acknowledge the APC waiver for contributing to this invited review article.

Conflicts of Interest: The authors declare no conflict of interest.

References

1. Fernandes, A.S.; do Nascimento, T.C.; Jacob-Lopes, E.; De Rosso, V.V.; Zepka, L.Q. Introductory Chapter: Carotenoids—A Brief Overview on Its Structure, Biosynthesis, Synthesis, and Applications. *Prog. Carotenoid Res.* **2018**, 1–16. [CrossRef]

2. Guedes, A.C.; Amaro, H.M.; Malcata, F.X. Microalgae as sources of carotenoids. *Mar. Drugs* **2011**, *9*, 625–644. [CrossRef]

3. Roberts, J.E.; Dennison, J. The Photobiology of Lutein and Zeaxanthin in the Eye. *J. Ophthalmol.* **2015**, *2015*, 687173. [CrossRef] [PubMed]

4. Marino, T.; Casella, P.; Sangiorgio, P.; Verardi, A.; Ferraro, A.; Hristoforou, E.; Molino, A.; Musmarra, D. Natural beta-carotene: A microalgae derivate for nutraceutical applications. *Chem. Eng. Trans.* **2020**, *79*, 103–108. [CrossRef]

5. Lin, J.H.; Lee, D.J.; Chang, J.S. Lutein production from biomass: Marigold flowers versus microalgae. *Bioresour. Technol.* **2015**, *184*, 421–428. [CrossRef]

6. Khan, M.I.; Shin, J.H.; Kim, J.D. The promising future of microalgae: Current status, challenges, and optimization of a sustainable and renewable industry for biofuels, feed, and other products. *Microb. Cell Fact.* **2018**, *17*, 1–21. [CrossRef]

7. Saha, S.K.; Murray, P. Exploitation of microalgae species for nutraceutical purposes: Cultivation aspects. *Fermentation* **2018**, *4*, 46. [CrossRef]

8. Bilal, M.; Rasheed, T.; Ahmed, I.; Iqbal, H.M.N. High-value compounds from microalgae with industrial exploitability—A review. *Front. Biosci. Sch.* **2017**, *9*, 319–342. [CrossRef]

9. Saha, S.K.; Mchugh, E.; Murray, P.; Walsh, D.J. Microalgae as a source of nutraceuticals. In *Phycotoxins: Chemistry and Biochemistry*, 2nd ed.; Botana, L.M., Alfonso, A., Eds.; JohnWiley & Sons Ltd.: Chichester, UK, 2015; pp. 255–292.

10. Roberts, R.L.; Green, J.; Lewis, B. Lutein and zeaxanthin in eye and skin health. *Clin. Dermatol.* **2009**, *27*, 195–201. [CrossRef]

11. Miller, D.L.; Papayannopoulos, I.A.; Styles, J.; Bobin, S.A.; Lin, Y.Y.; Biemann, K.; Iqbal, K. Peptide Compositions of the Cerebrovascular and Senile Plaque Core Amyloid Deposits of Alzheimer's Disease. *Arch. Biochem. Biophys.* **1993**, *301*, 41–52. [CrossRef]

12. Kim, J.K.; Park, S.U. Letter to the editor: Current results on the potential health benefits of lutein. *EXCLI J.* **2016**, *15*, 308–314. [PubMed]

13. Giordano, E.; Quadro, L. Lutein, zeaxanthin and mammalian development: Metabolism, functions and implications for health. *Arch. Biochem. Biophys.* **2018**, *647*, 33–40. [CrossRef] [PubMed]

14. Fitzpatrick, E.; Dhawan, A. Scanning the scars: The utility of transient elastography in young children. *J. Pediatr. Gastroenterol. Nutr.* **2014**, *59*, 551. [CrossRef] [PubMed]

15. Eggersdorfer, M.; Wyss, A. Carotenoids in human nutrition and health. *Arch. Biochem. Biophys.* **2018**, *652*, 18–26. [CrossRef]

16. Ochoa Becerra, M.; Mojica Contreras, L.; Hsieh Lo, M.; Mateos Díaz, J.; Castillo Herrera, G. Lutein as a functional food ingredient: Stability and bioavailability. *J. Funct. Foods* **2020**, *66*, 103771. [CrossRef]

17. Takaichi, S. Carotenoids in algae: Distributions, biosyntheses and functions. *Mar. Drugs* **2011**, *9*, 1101–1118. [CrossRef]

18. Barredo, J.-L. Microbial Carotenoids from Bacteria and Microalgae. Methods and Protocols. *Methods Mol. Biol.* **2012**, *892*, 133–141. [CrossRef]

19. Gwak, Y.; Hwang, Y.S.; Wang, B.; Kim, M.; Jeong, J.; Lee, C.G.; Hu, Q.; Han, D.; Jin, E. Comparative analyses of lipidomes and transcriptomes reveal a concerted action of multiple defensive systems against photooxidative stress in *Haematococcus pluvialis*. *J. Exp. Bot.* **2014**, *65*, 4317–4334. [CrossRef]

20. Hunter, W.N. The non-mevalonate pathway of isoprenoid precursor biosynthesis. *J. Biol. Chem.* **2007**, *282*, 21573–21577. [CrossRef]

21. Kim, J.; Kim, M.; Lee, S.; Jin, E.S. Development of a *Chlorella vulgaris* mutant by chemical mutagenesis as a producer for natural violaxanthin. *Algal Res.* **2020**, *46*, 101790. [CrossRef]

22. Lee, P.C.; Schmidt-Dannert, C. Metabolic engineering towards biotechnological production of carotenoids in microorganisms. *Appl. Microbiol. Biotechnol.* **2002**, *60*, 1–11. [CrossRef] [PubMed]

23. Fan, X.D.; Hou, Y.; Huang, X.X.; Qiu, T.Q.; Jiang, J.G. Ultrasound-enhanced subcritical CO_2 extraction of lutein from *Chlorella pyrenoidosa*. *J. Agric. Food Chem.* **2015**, *63*, 4597–4605. [CrossRef] [PubMed]

24. Yildirim, A.; Akgün, İ.H.; Conk Dalay, M. Converted carotenoid production in *Dunaliella salina* by using cyclization inhibitors 2-methylimidazole and 3-amino-1,2,4-triazole. *Turkish J. Biol.* **2017**, *41*, 213–219. [CrossRef]

25. Liang, M.H.; Hao, Y.F.; Li, Y.M.; Liang, Y.J.; Jiang, J.G. Inhibiting Lycopene Cyclases to Accumulate Lycopene in High β-Carotene-Accumulating *Dunaliella bardawil*. *Food Bioprocess Technol.* **2016**, *9*, 1002–1009. [CrossRef]

26. Ishikawa, E.; Abe, H. Lycopene accumulation and cyclic carotenoid deficiency in heterotrophic *Chlorella* treated with nicotine. *J. Ind. Microbiol. Biotechnol.* **2004**, *31*, 585–589. [CrossRef]

27. Fazeli, M.R.; Tofighi, H.; Madadkar-Sobhani, A.; Shahverdi, A.R.; Nejad-Sattari, T.; Sako, M.; Jamalifar, H. Nicotine inhibition of lycopene cyclase enhances accumulation of carotenoid intermediates by *Dunaliella salina* CCAP 19/18. *Eur. J. Phycol.* **2009**, *44*, 215–220. [CrossRef]

28. Cordero, B.F.; Obraztsova, I.; Couso, I.; Leon, R.; Vargas, M.A.; Rodriguez, H. Enhancement of lutein production in *Chlorella sorokiniana* (chorophyta) by improvement of culture conditions and random mutagenesis. *Mar. Drugs* **2011**, *9*, 1607–1624. [CrossRef]

29. Kamath, B.S.; Vidhyavathi, R.; Sarada, R.; Ravishankar, G.A. Enhancement of carotenoids by mutation and stress induced carotenogenic genes in *Haematococcus pluvialis* mutants. *Bioresour. Technol.* **2008**, *99*, 8667–8673. [CrossRef]

30. Huang, W.; Lin, Y.; He, M.; Gong, Y.; Huang, J. Induced high-yield production of zeaxanthin, lutein, and β-carotene by a mutant of *Chlorella zofingiensis*. *J. Agric. Food Chem.* **2018**, *66*, 891–897. [CrossRef]

31. Gimpel, J.A.; Henríquez, V.; Mayfield, S.P. In metabolic engineering of eukaryotic microalgae: Potential and challenges come with great diversity. *Front. Microbiol.* **2015**, *6*, 1–14. [CrossRef]

32. Coll, J.M. Methodologies for transferring DNA into eukaryotic microalgae. *Span. J. Agric. Res.* **2006**, *4*, 316–330. [CrossRef]

33. Terashima, M.; Specht, M.; Hippler, M. The chloroplast proteome: A survey from the *Chlamydomonas reinhardtii* perspective with a focus on distinctive features. *Curr. Genet.* **2011**, *57*, 151–168. [CrossRef] [PubMed]

34. Neu, T.R.; Lawrence, J.R. Investigation of Microbial Biofilm Structure by Laser Scanning Microscopy. *Adv. Biochem. Eng. Biotechnol.* **2014**, *123*, 127–141. [CrossRef]

35. Lizzul, A.M.; Lekuona-Amundarain, A.; Purton, S.; Campos, L.C. Characterization of *Chlorella sorokiniana*, UTEX 1230. *Biology* **2018**, *7*, 25. [CrossRef] [PubMed]

36. Cordero, B.F.; Couso, I.; León, R.; Rodríguez, H.; Vargas, M.Á. Enhancement of carotenoids biosynthesis in *Chlamydomonas reinhardtii* by nuclear transformation using a phytoene synthase gene isolated from *Chlorella zofingiensis*. *Appl. Microbiol. Biotechnol.* **2011**, *91*, 341–351. [CrossRef] [PubMed]

37. Liu, J.; Gerken, H.; Huang, J.; Chen, F. Engineering of an endogenous phytoene desaturase gene as a dominant selectable marker for *Chlamydomonas reinhardtii* transformation and enhanced biosynthesis of carotenoids. *Process Biochem.* **2013**, *48*, 788–795. [CrossRef]

38. Liu, J.; Sun, Z.; Gerken, H.; Huang, J.; Jiang, Y.; Chen, F. Genetic engineering of the green alga *Chlorella zofingiensis*: A modified norflurazon-resistant phytoene desaturase gene as a dominant selectable marker. *Appl. Microbiol. Biotechnol.* **2014**, *98*, 5069–5079. [CrossRef]

39. Rathod, J.P.; Vira, C.; Lali, A.M.; Prakash, G. Metabolic Engineering of *Chlamydomonas reinhardtii* for Enhanced β-Carotene and Lutein Production. *Appl. Biochem. Biotechnol.* **2020**, *190*, 1457–1469. [CrossRef] [PubMed]

40. Majer, E.; Llorente, B.; Rodríguez-Concepción, M.; Daròs, J.A. Rewiring carotenoid biosynthesis in plants using a viral vector. *Sci. Rep.* **2017**, *7*, 41645. [CrossRef] [PubMed]

41. Bogacz-Radomska, L.; Harasym, J. β-Carotene-properties and production methods. *Food Qual. Saf.* **2018**, *2*, 69–74. [CrossRef]

42. El-Gawad, E.A.A.; Wang, H.P.; Yao, H. Diet Supplemented With Synthetic Carotenoids: Effects on Growth Performance and Biochemical and Immunological Parameters of Yellow Perch (*Perca flavescens*). *Front. Physiol.* **2019**, *10*, 1–13. [CrossRef] [PubMed]

43. Khachik, F.; Chang, A.N. Total synthesis of (3R,3′R,6′R)-lutein and its stereoisomers. *J. Org. Chem.* **2009**, *74*, 3875–3885. [CrossRef] [PubMed]

44. Hou, M.; Wang, R.; Wu, X.; Zhang, Y.; Ge, J.; Liu, Z. Synthesis of lutein esters by using a reusable lipase-Pluronic conjugate as the catalyst. *Catal. Lett.* **2015**, *145*, 1825–1829. [CrossRef]

45. Novoveská, L.; Ross, M.E.; Stanley, M.S.; Pradelles, R.; Wasiolek, V.; Sassi, J.F. Microalgal carotenoids: A review of production, current markets, regulations, and future direction. *Mar. Drugs* **2019**, *17*, 640. [CrossRef] [PubMed]

46. Ho, S.H.; Chan, M.C.; Liu, C.C.; Chen, C.Y.; Lee, W.L.; Lee, D.J.; Chang, J.S. Enhancing lutein productivity of an indigenous microalga *Scenedesmus obliquus* FSP-3 using light-related strategies. *Bioresour. Technol.* **2014**, *152*, 275–282. [CrossRef]

47. Ho, S.H.; Xie, Y.; Chan, M.C.; Liu, C.C.; Chen, C.Y.; Lee, D.J.; Huang, C.C.; Chang, J.S. Effects of nitrogen source availability and bioreactor operating strategies on lutein production with *Scenedesmus obliquus* FSP-3. *Bioresour. Technol.* **2015**, *184*, 131–138. [CrossRef]

48. Zhao, X.; Ma, R.; Liu, X.; Ho, S.H.; Xie, Y.; Chen, J. Strategies related to light quality and temperature to improve lutein production of marine microalga *Chlamydomonas* sp. *Bioprocess Biosyst. Eng.* **2019**, *42*, 435–443. [CrossRef]

49. Schüler, L.M.; Santos, T.; Pereira, H.; Duarte, P.; Katkam, N.G.; Florindo, C.; Schulze, P.S.C.; Barreira, L.; Varela, J.C.S. Improved production of lutein and β-carotene by thermal and light intensity upshifts in the marine microalga *Tetraselmis* sp. CTP4. *Algal Res.* **2020**, *45*, 101732. [CrossRef]

50. Ma, R.; Zhao, X.; Xie, Y.; Ho, S.H.; Chen, J. Enhancing lutein productivity of *Chlamydomonas* sp. via high-intensity light exposure with corresponding carotenogenic genes expression profiles. *Bioresour. Technol.* **2019**, *275*, 416–420. [CrossRef]

51. Chen, W.C.; Hsu, Y.C.; Chang, J.S.; Ho, S.H.; Wang, L.F.; Wei, Y.H. Enhancing production of lutein by a mixotrophic cultivation system using microalga *Scenedesmus obliquus* CWL-1. *Bioresour. Technol.* **2019**, *291*, 121891. [CrossRef]

52. Flórez-Miranda, L.; Cañizares-Villanueva, R.O.; Melchy-Antonio, O.; Martínez-Jerónimo, F.; Flores-Ortíz, C.M. Two stage heterotrophy/photoinduction culture of *Scenedesmus incrassatulus*: Potential for lutein production. *J. Biotechnol.* **2017**, *262*, 67–74. [CrossRef]

53. Xie, Y.; Li, J.; Ma, R.; Ho, S.H.; Shi, X.; Liu, L.; Chen, J. Bioprocess operation strategies with mixotrophy/photoinduction to enhance lutein production of microalga *Chlorella sorokiniana* FZU60. *Bioresour. Technol.* **2019**, *290*, 121798. [CrossRef] [PubMed]

54. Shi, X.M.; Jiang, Y.; Chen, F. High-yield production of lutein by the green microalga *Chlorella protothecoides* in heterotrophic fed-batch culture. *Biotechnol. Prog.* **2002**, *18*, 723–727. [CrossRef] [PubMed]

55. Shinde, S.; Lele, S. Statistical media optimization for lutein production from microalgae *Auxenochlorella protothecoides* SAG 211-7A. *Int. J. Adv. Biotechnol. Res.* **2010**, *1*, 104–114.

56. Borowitzka, M.A.; Borowitzka, L.J.; Kessly, D. Effects of salinity increase on carotenoid accumulation in the green alga *Dunaliella salina*. *J. Appl. Phycol.* **1990**, *2*, 111–119. [CrossRef]

57. Blanco, A.M.; Moreno, J.; Del Campo, J.A.; Rivas, J.; Guerrero, M.G. Outdoor cultivation of lutein-rich cells of *Muriellopsis* sp. in open ponds. *Appl. Microbiol. Biotechnol.* **2007**, *73*, 1259–1266. [CrossRef]

58. Poojary, M.M.; Barba, F.J.; Aliakbarian, B.; Donsì, F.; Pataro, G.; Dias, D.A.; Juliano, P. Innovative alternative technologies to extract carotenoids from microalgae and seaweeds. *Mar. Drugs* **2016**, *14*, 214. [CrossRef]

59. Kumar, S.P.J.; Kumar, G.V.; Dash, A.; Scholz, P.; Banerjee, R. Sustainable green solvents and techniques for lipid extraction from microalgae: A review. *Algal Res.* **2017**, *21*, 138–147. [CrossRef]

60. Macías-Sánchez, M.D.; Fernandez-Sevilla, J.M.; Fernández, F.G.A.; García, M.C.C.; Grima, E.M. Supercritical fluid extraction of carotenoids from *Scenedesmus almeriensis*. *Food Chem.* **2010**, *123*, 928–935. [CrossRef]

61. Araya, B.; Gouveia, L.; Nobre, B.; Reis, A.; Chamy, R.; Poirrier, P. Evaluation of the simultaneous production of lutein and lipids using a vertical alveolar panel bioreactor for three *Chlorella* species. *Algal Res.* **2014**, *6*, 218–222. [CrossRef]

62. Chen, C.Y.; Jesisca; Hsieh, C.; Lee, D.J.; Chang, C.H.; Chang, J.S. Production, extraction and stabilization of lutein from microalga *Chlorella sorokiniana* MB-1. *Bioresour. Technol.* **2016**, *200*, 500–505. [CrossRef] [PubMed]

63. Nonomura, A.M. Process for Producing a Naturally-Derived Carotene/Oil Composition by Direct Extraction from Algae. U.S. Patent 4,680,314A, 14 July 1987.

64. Low, K.L.; Idris, A.; Mohd Yusof, N. Novel protocol optimized for microalgae lutein used as food additives. *Food Chem.* **2020**, *307*, 125631. [CrossRef]

65. Cerón, M.C.; Campos, I.; Sánchez, J.F.; Acién, F.G.; Molina, E.; Fernández-Sevilla, J.M. Recovery of lutein from microalgae biomass: Development of a process for *Scenedesmus almeriensis* biomass. *J. Agric. Food Chem.* **2008**, *56*, 11761–11766. [CrossRef]

66. Gong, M.; Li, X.; Bassi, A. Investigation of simultaneous lutein and lipid extraction from wet microalgae using Nile Red as solvatochromic shift probe. *J. Appl. Phycol.* **2018**, *30*, 1617–1627. [CrossRef]

67. Li, H.B.; Chen, F. Preparative isolation and purification of astaxanthin from the microalga *Chlorococcum* sp. by high-speed counter-current chromatography. *J. Chromatogr. A* **2001**, *925*, 133–137. [CrossRef]

68. Yen, H.W.; Chiang, W.C.; Sun, C.H. Supercritical fluid extraction of lutein from *Scenedesmus* cultured in an autotrophical photobioreactor. *J. Taiwan Inst. Chem. Eng.* **2012**, *43*, 53–57. [CrossRef]

69. Miguel, F.; Martín, A.; Mattea, F.; Cocero, M.J. Precipitation of lutein and co-precipitation of lutein and poly-lactic acid with the supercritical anti-solvent process. *Chem. Eng. Process. Process Intensif.* **2008**, *47*, 1594–1602. [CrossRef]

70. Di Sanzo, G.; Mehariya, S.; Martino, M.; Larocca, V.; Casella, P.; Chianese, S.; Musmarra, D.; Balducchi, R.; Molino, A. Supercritical carbon dioxide extraction of astaxanthin, lutein, and fatty acids from *Haematococcus pluvialis* microalgae. *Mar. Drugs* **2018**, *16*, 334. [CrossRef]

71. Mehariya, S.; Iovine, A.; Di Sanzo, G.; Larocca, V.; Martino, M.; Leone, G.P.; Casella, P.; Karatza, D.; Marino, T.; Musmarra, D.; et al. Supercritical fluid extraction of lutein from *Scenedesmus almeriensis*. *Molecules* **2019**, *24*, 1324. [CrossRef]

72. Ruen-Ngam, D.; Shotipruk, A.; Pavasant, P.; Machmudah, S.; Goto, M. Selective extraction of lutein from alcohol treated *Chlorella vulgaris* by supercritical CO_2. *Chem. Eng. Technol.* **2012**, *35*, 255–260. [CrossRef]

73. Zhengyun, W.; Wu, S.; Xianming, S. Supercritical fluid extraction and determination of lutein in heterotrophically cultivated *Chlorella pyrenoidosa*. *J. Food Process Eng.* **2007**, *30*, 174–185. [CrossRef]

74. Ishida, B.K.; Chapman, M.H. Carotenoid Extraction from Plants Using a Novel, Environmentally Friendly Solvent. *J. Agric. Food Chem.* **2009**, *57*, 1051–1059. [CrossRef]

75. Ma, R.; Zhao, X.; Ho, S.H.; Shi, X.; Liu, L.; Xie, Y.; Chen, J.; Lu, Y. Co-production of lutein and fatty acid in microalga *Chlamydomonas* sp. JSC4 in response to different temperatures with gene expression profiles. *Algal Res.* **2020**, *47*, 101821. [CrossRef]

76. Xie, Y.; Lu, K.; Zhao, X.; Ma, R.; Chen, J.; Ho, S.H. Manipulating Nutritional Conditions and Salinity-Gradient Stress for Enhanced Lutein Production in Marine Microalga *Chlamydomonas* sp. *Biotechnol. J.* **2019**, *14*, 1–28. [CrossRef] [PubMed]

77. Garbayo, I.; Cuaresma, M.; Vílchez, C.; Vega, J.M. Effect of abiotic stress on the production of lutein and β-carotene by *Chlamydomonas acidophila*. *Process Biochem.* **2008**, *43*, 1158–1161. [CrossRef]

78. Gayathri, S.; Radhika, S.R.R.; Suman, T.Y.; Aranganathan, L. Ultrasound-assisted microextraction of β, ε-carotene-3, 3′-diol (lutein) from marine microalgae *Chlorella salina*: Effect of different extraction parameters. *Biomass Convers. Biorefinery* **2018**, *8*, 791–797. [CrossRef]

79. Serejo, M.L.; Posadas, E.; Boncz, M.A.; Blanco, S.; García-Encina, P.; Muñoz, R. Influence of biogas flow rate on biomass composition during the optimization of biogas upgrading in microalgal-bacterial processes. *Environ. Sci. Technol.* **2015**, *49*, 3228–3236. [CrossRef]

80. Del Campo, J.A.; Rodríguez, H.; Moreno, J.; Vargas, M.Á.; Rivas, J.; Guerrero, M.G. Lutein production by *Muriellopsis* sp. in an outdoor tubular photobioreactor. *J. Biotechnol.* **2001**, *85*, 289–295. [CrossRef]

81. Dineshkumar, R.; Dhanarajan, G.; Dash, S.K.; Sen, R. An advanced hybrid medium optimization strategy for the enhanced productivity of lutein in *Chlorella minutissima*. *Algal Res.* **2015**, *7*, 24–32. [CrossRef]

82. Shi, X.M.; Liu, H.J.; Zhang, X.W.; Chen, F. Production of biomass and lutein by *Chlorella protothecoides* at various glucose concentrations in heterotrophic cultures. *Process Biochem.* **1999**, *34*, 341–347. [CrossRef]

83. Chen, C.Y.; Liu, C.C. Optimization of lutein production with a two-stage mixotrophic cultivation system with *Chlorella sorokiniana* MB-1. *Bioresour. Technol.* **2018**, *262*, 74–79. [CrossRef] [PubMed]

84. Deenu, A.; Naruenartwongsakul, S.; Kim, S.M. Optimization and economic evaluation of ultrasound extraction of lutein from *Chlorella vulgaris*. *Biotechnol. Bioprocess Eng.* **2013**, *18*, 1151–1162. [CrossRef]

85. Xin, C.; Addy, M.M.; Zhao, J.; Cheng, Y.; Cheng, S.; Mu, D.; Liu, Y.; Ding, R.; Chen, P.; Ruan, R. Comprehensive techno-economic analysis of wastewater-based algal biofuel production: A case study. *Bioresour. Technol.* **2016**, *211*, 584–593. [CrossRef] [PubMed]

86. Del Campo, J.A.; Rodríguez, H.; Moreno, J.; Vargas, M.Á.; Rivas, J.; Guerrero, M.G. Accumulation of astaxanthin and lutein in *Chlorella zofingiensis* (Chlorophyta). *Appl. Microbiol. Biotechnol.* **2004**, *64*, 848–854. [CrossRef]

87. Xie, Y.; Zhao, X.; Chen, J.; Yang, X.; Ho, S.H.; Wang, B.; Chang, J.S.; Shen, Y. Enhancing cell growth and lutein productivity of *Desmodesmus* sp. F51 by optimal utilization of inorganic carbon sources and ammonium salt. *Bioresour. Technol.* **2017**, *244*, 664–671. [CrossRef]

88. Del Campo, J.A.; Moreno, J.; Rodríguez, H.; Angeles Vargas, M.; Rivas, J.; Guerrero, M.G. Carotenoid content of chlorophycean microalgae: Factors determining lutein accumulation in *Muriellopsis* sp. (Chlorophyta). *J. Biotechnol.* **2000**, *76*, 51–59. [CrossRef]

89. Ram, S.; Paliwal, C.; Mishra, S. Growth medium and nitrogen stress sparked biochemical and carotenogenic alterations in *Scenedesmus* sp. CCNM 1028. *Bioresour. Technol. Rep.* **2019**, *7*, 100194. [CrossRef]

90. Molino, A.; Mehariya, S.; Karatza, D.; Chianese, S.; Iovine, A.; Casella, P.; Marino, T.; Musmarra, D. Bench-scale cultivation of microalgae *Scenedesmus almeriensis* for CO_2 capture and lutein production. *Energies* **2019**, *12*, 2806. [CrossRef]

91. Sánchez, J.F.; Fernández, J.M.; Acién, F.G.; Rueda, A.; Pérez-Parra, J.; Molina, E. Influence of culture conditions on the productivity and lutein content of the new strain *Scenedesmus almeriensis*. *Process Biochem.* **2008**, *43*, 398–405. [CrossRef]

92. Lutein Market to Reach USD 454.8 Million by 2026|Reports And Data. Available online: https://www.prnewswire. com/news-releases/lutein-market-to-reach-usd-454-8-million-by-2026--reports-and-data-300941985.html (accessed on 4 September 2020).

93. Ambati, R.R.; Gogisetty, D.; Aswathanarayana, R.G.; Ravi, S.; Bikkina, P.N.; Bo, L.; Yuepeng, S. Industrial potential of carotenoid pigments from microalgae: Current trends and future prospects. *Crit. Rev. Food Sci. Nutr.* **2019**, *59*, 1880–1902. [CrossRef]

94. Gong, M.; Bassi, A. Carotenoids from microalgae: A review of recent developments. *Biotechnol. Adv.* **2016**, *34*, 1396–1412. [CrossRef] [PubMed]

applied
sciences

Article

Pigment and Fatty Acid Production under Different Light Qualities in the Diatom *Phaeodactylum tricornutum*

Bernardo Duarte [1,2,*] , Eduardo Feijão [1], Johannes W. Goessling [3], Isabel Caçador [1,2] and Ana Rita Matos [2,4]

1   MARE—Marine and Environmental Sciences Centre, Faculdade de Ciências da Universidade de Lisboa, Campo Grande, 1749-016 Lisboa, Portugal; emfeijao@fc.ul.pt (E.F.); micacador@fc.ul.pt (I.C.)
2   Departamento de Biologia Vegetal, Faculdade de Ciências da Universidade de Lisboa, Campo Grande, 1749-016 Lisboa, Portugal; armatos@fc.ul.pt
3   International Iberian Nanotechnology Laboratory, 4715-330 Braga, Portugal; johannes.goessling@inl.int
4   Plant Functional Genomics Group, BioISI—Biosystems and Integrative Sciences Institute, Faculdade de Ciências da Universidade de Lisboa, Campo Grande, 1749-016 Lisboa, Portugal
*   Correspondence: baduarte@fc.ul.pt

Featured Application: light emitting diodes (LED) illumination with different wavelengths can modulate diatom ca-rotenoid and fatty acid production.

Abstract: Diatoms are microscopic biorefineries producing value-added molecules, including unique pigments, triglycerides (TAGs) and long-chain polyunsaturated fatty acids (LC-PUFAs), with potential implications in aquaculture feeding and the food or biofuel industries. These molecules are utilized in vivo for energy harvesting from sunlight to drive photosynthesis and as photosynthetic storage products, respectively. In the present paper, we evaluate the effect of narrow-band spectral illumination on carotenoid, LC-PUFAs and TAG contents in the model diatom *Phaeodactylum tricornutum*. Shorter wavelengths in the blue spectral range resulted in higher production of total fatty acids, namely saturated TAGs. Longer wavelengths in the red spectral range increased the cell's content in Hexadecatrienoic acid (HTA) and Eicosapentaenoic acid (EPA). Red wavelengths induced higher production of photoprotective carotenoids, namely fucoxanthin. In combination, the results demonstrate how diatom value-added molecule production can be modulated by spectral light control during the growth. How diatoms could use such mechanisms to regulate efficient light absorption and cell buoyancy in the open ocean is discussed.

Keywords: *Phaedactylum tricornutum*; photochemistry; fucoxanthin; single wavelength LEDs

Citation: Duarte, B.; Feijão, E.; Goessling, J.W.; Caçador, I.; Matos, A.R. Pigment and Fatty Acid Production under Different Light Qualities in the Diatom *Phaeodactylum tricornutum. Appl. Sci.* **2021**, *11*, 2550. https://doi.org/10.3390/app11062550

Academic Editor: Guillaume Pierre

Received: 24 February 2021
Accepted: 8 March 2021
Published: 12 March 2021

Publisher's Note: MDPI stays neutral with regard to jurisdictional claims in published maps and institutional affiliations.

1. Introduction

Diatoms (Bacillariophyta) are among the most abundant phytoplankton species on Earth [1]. They play vital roles as primary producers in aquatic food webs, being responsible for half of the organic materials in the ocean and up to 20% of the Earth's oxygen produced by photosynthesis [1]. Their high productivity has attracted scientific study related to their biotechnological potential, e.g., as biorefineries [2]. Diatoms have thereby mainly been exploited by the nutraceutical, fuel and aquaculture sectors, emerging as novel bioresources for the production of bioenergy, food and aquaculture feed, and as supporter of wastewater bioremediation [3]. Besides the most recent approaches to improve microalgae value-added yield production, using genetic engineering and molecular biology approaches [4], the manipulation of culture conditions (e.g., the light availability) has also been investigated [5]. The unique products of diatoms originate from their photosynthetic metabolism. Fucoxanthin (Fx) is a primary carotenoid with expanded light absorption in the cyan and green spectral range of light. It has drawn attention due to its antioxidative properties, as potential antiobesity, and potential anticancer compound as well as some suggested effects to mitigate Alzheimer's disease [6–8]. Today, the main commercially available source of Fx is brown seaweeds (Stramenopiles), a phylogenetic group to which the diatoms also belong [9]. Diatoms store photosynthetic energy foremost in form of fatty acids, with high proportions of polyunsaturated fatty acids (PUFAs), namely ω3 fatty acids such as eicosapentaenoic acid (20:5, EPA) (Dunstan et al., 1993). Additionally, diatoms show high amounts of storage lipids (triacylglycerols, TAG) [10], with high nutritional value [11] and are useful for aquaculture feeding and potentially biofuel production [12].

Diatoms require light for their growth, as their metabolism bases on photosynthesis as primary energy source [13]. Some common cultivation systems involve light emitting diodes (LEDs) to support algal growth in cell cultures at the maximum light absorption of the main photosynthetic pigments, in the blue and red spectral range of light at ca. $\lambda \approx 450$ nm and $\lambda \approx 650$ nm, respectively [5]. The narrow-band wavelengths of LEDs can also be applied to test physiological implications of particular spectral parts of light and metabolite production under more environmental-like conditions, where light intensity as well as light spectral composition are affected by the wavelengths dependent light attenuation characteristics of water [14]. Individuals near the surface might thereby receive significantly higher amounts of light, compared to cells located at subsurface layers in the water column. In addition, light wavelengths of higher energy in the blue spectral range can penetrate deeper in the absence of dissolved organic matter and microscale particles [15]. In consequence, individuals of the same species can be projected to significant differences of available light during their lifetime. The different wavelengths to which diatoms are subjected in their natural environment and at different life stages are key factors shaping the metabolic activity of the cells [16]. The presence of blue and red light photoreceptors, i.e., cryptochromes and aureochromes, and phytochromes, respectively, allows diatoms to sense their position in the water column and to perform morphological and physiological alterations to adapt to changes in environmental conditions [17,18].

In the present work, we studied the effects of narrow-band illumination with blue and red light LEDs compared to red-green-blue (RGB) LEDs upon photo-pigment, lipid and fatty acid production in the diatom *Phaeodactylum tricornutum*. Different pulse amplitude modulated (PAM) chlorophyll fluorescence techniques were used to determine the energy flux and dissipation along the photosynthetic transport chain under these illumination conditions. *Phaeodactylum tricornutum* is a well characterized model diatom, which can be cultivated in several culture media [2]. It is known for a variety of marketable products and today commercially viable for large-scale cultivation in some cases [2]. The data presented here suggest that spectral illumination conditions can modulate pigment concentrations and the composition of fatty acids, paving the way for optimized production of these high-value products in *P. tricornutum*. We speculate that some can regulate their position in the water column by a fine-tuned interplay of light absorption by photo-pigments and the composition and quantity of photosynthetic storage lipids to control the cell buoyancy.

2. Materials and Methods

2.1. Culture Conditions

Phaeodactylum tricornutum Bohlin (Bacillariophyceae; strain IO 108–01, Instituto Português do Mar e da Atmosfera (IPMA)) axenic cell cultures were placed to grow in f/2 medium [19], under constant aeration in a phytoclimatic chamber (FytoScope FS 130—RGBIR, Photon Systems Instruments, Czech Republic), at 18 °C, programmed with a 14/10 h day/night photoperiod using a sinusoidal function provided by the manufacturer to mimic sunrise and sunset, and light intensity at noon, set to replicate a natural light environment, with a maximum light intensity of 80 μmol photons m^{-2} s^{-1} at solar noon. [20]. Cultures were daily inspected visually under the microscope. Culture trials under the different light regimes were conducted according to the Organization for Economic Cooperation and Development (OECD) recommendations for algae assays [21], with minor adaptations, and the suggested initial cell density for microalgae cells with comparable dimensions to *P. tricornutum* (initial cell density = 2.7×10^5 cells mL^{-1}). According to OECD guidelines, carbon dioxide concentrations were maintained through constant aeration of the cultures with ambient air. All manipulations were executed within a laminar flow hood chamber, ensuring standard aseptic conditions.

Light conditions were adjusted using the LED panel of the FytoScope FS 130—RGBIR. Full light treatment (denoted as RGB LED hereafter) was provided using a combination of RGB LEDs in the molar proportion 1:1:1 (Red λ = 627 nm/Green λ = 530 nm/Blue λ = 470 nm). Red light treatment (denoted as Red LED hereafter), was preformed using only the Red LEDs (λ = 627 nm). Blue light treatment (denoted as Blue LED hereafter), was provided using only the Blue LEDs (λ = 470 nm). All light intensities were adjusted in all light conditions to achieve a maximum photosynthetically active radiation (PAR) of 80 μmol photons m^{-2} s^{-1} at solar noon, at the culture flasks level. After inoculation cultures were immediately exposed to the targeted light treatment, and the experiment lasted for 96 h.

2.2. Diatom Cell Density Measurements and Pellet Collection

Phaeodactylum tricornutum cells (1 mL volume sample) were counted using a Neubauer improved counting chamber, coupled with an Olympus BX50 (Tokyo, Japan) inverted microscope, at 400-x magnification. At the end of the exposure trials, cells were harvested for biochemical analysis by centrifugation at 4000$\times$ *g* for 15 min at 4 °C and the pellets were frozen in liquid nitrogen and stored at −80 °C. Five biological replicates for all tested conditions were considered for total fatty acid analysis and pigment analysis and three replicates were considered for triacylglyceride (TAG) quantification.

2.3. Chlorophyll a Pulse Amplitude Modulated Fluorometry

At the end of the experimental period, and before cell harvesting, 1 mL of each replicate culture was used for bio-optical assessment, using chlorophyll-a pulse amplitude modulated (PAM) fluorometry (FluorPen FP100, Photo System Instruments, Brno, Czech Republic). Cell culture subsamples for bio-optical assessment were acclimated for 15 min in the dark and chlorophyll transient light curves were generated using the preprogrammed OJIP (fluorescence rise through four phases called O, J, I and P) protocol [22]. Rapid light curves (RLC) were generated using the preprogrammed LC1. The parameters determined and calculated by the software from this analysis are shown in Table 1 [23,24].

Table 1. Summary of fluorometric analysis parameters and their description.

Variable	Description
Φ PSII	Photosystem II (PSII) maximum quantum yield (F_v/F_m, where F_v is the variable fluorescence and F_m is the maximum fluorescence).
α	Photosynthetic efficiency, corresponding to the initial slope of the relative electron transport rate (rETR) versus photosynthetic photon flux density (PPFD) curve.
Q_{phar} α	Photosynthetic efficiency, corresponding to the initial slope of the relative electron transport rate (rETR) versus pigment weighted light absorption.
rETR	Relative electron transport rate, obtained from applying the equation: Φ PSII $\times$ PPFD $\times$ 0.5 (factor for correcting for the energy generated only at the PSII side, assuming each photosystem absorbs 50% of the incoming energy).
ETR$_{max}$	Maximum ETR having as basis the PPFD.
Q_{phar} ETR$_{max}$	Maximum ETR calculated using the pigment weighted light absorption.
AOECS	Active oxygen evolving complexes at the PSII donor side.
ABS/CS	Absorbed energy flux per cross-section.
TR/CS	Trapped energy flux per cross-section.
ET/CS	Electron transport energy flux per cross-section.
DI/CS	Dissipated energy flux per cross-section.
RC/CS	Number of available reaction centers per cross-section.

2.4. Pigment Profiles

Pigments were extracted from sample pellets with 100% cold acetone and maintained in a cold ultra-sound bath for 2 min, to ensure complete disaggregation of the cell material. Extraction proceeded in the dark at $-20\,°C$ for 24 h, to prevent pigment degradation [20,25,26]. Following centrifugation ($4000\times g$ for 15 min at $4\,°C$), supernatants were analyzed using a dual beam spectrophotometer. Absorbance spectrums from 350 nm to 750 nm (0.5 nm steps) were then introduced in the Gauss-peak spectra (GPS) fitting library, using SigmaPlot software. Pigment analysis was employed using the a gaussian peak deconvolution algorithm [27], enabling the detection of Chlorophyll a and c, Pheophytin a, β-carotene, Fx, Diadinoxanthin (DD), and Diatoxanthin (DT).

2.5. Absorption Spectra

To compare whether differences in light color responses were due to total light absorbed or due to light quality effects, a pigment weighted light absorption (Q_{Phar}) was calculated [28] with modification [29]. Q_{Phar} (λ) correspond to the amount and proportion of light absorbed by the cells at different wavelengths. The specific in vitro absorption coefficients of the *P. tricornutum* cultures (*Pt*) were reconstructed according to:

$$a_{Pt}(\lambda) = \sum_{i}^{n} a_i(\lambda) \times C_i$$

where $a_i(\lambda)$ is the concentration specific absorption coefficient obtained from the literature for each pigment at each incident wavelength [30] and C_i is the pigment concentration of the culture. For extraction of the correct absorption coefficient the predominant wavelengths of each of the LED light quality was used. The pigment weighted light absorption Q_{phar} is obtained from:

$$Q_{phar} = Q(\lambda) - \left[Q(\lambda) \times e^{-a_{Pt}(\lambda)} \right]$$

where $Q(\lambda)$ is the incident PAR in $\mu mol\ m^{-2}\ s^{-1}$ [28]. The pigments used for spectral absorption reconstruction were chlorophyll a, chlorophyll c, pheophytin a, β-carotene, Fx, DD and DT.

2.6. Total Fatty Acids and TAG Profiles

Cell pellets for total fatty acid analysis were submitted to direct transesterification with freshly prepared methanol sulfuric acid (97.5:2.5, v/v) at $70\,°C$ for 60 min [31]. Subsequently, fatty acids methyl esters (FAMEs) were recovered using petroleum ether and the solvent evaporated under a constant N_2 flow in a dry bath at $30\,°C$ [20,32]. FAMEs were resuspended in hexane and 1 µL was injected in a gas chromatograph (Varian 430-GC gas chromatograph, Middelburg, The Netherlands), equipped with a hydrogen flame ionization detector set at $300\,°C$. The temperature of the injector was set to $270\,°C$, with a split ratio of 50. The fused-silica capillary column (50 m $\times$ 0.25 mm; WCOT Fused Silica, CP-Sil 88 for FAME; Varian, Middelburg, The Netherlands) was maintained at a constant N_2 flow of 2.0 mL min^{-1} and the oven set at $190\,°C$. Fatty acids identification was achieved by comparison of retention times with standards (Sigma-Aldrich, St. Louis, MO, USA), and chromatograms analyzed by the peak surface method, using the Galaxy software (Varian, Inc., Palo Alto, CA, USA). The internal standard used was pentadecanoic acid (15:0) to identify losses during preparation.

Two indexes are commonly used to infer and predict the potential health benefits associated with the ingestion of a certain food: indexes of atherogenicity (*IA*) and thrombogenicity (*IT*) [33]:

$$IA = \frac{4 \times C14:0 + C16:0}{\sum MUFA + \sum PUFA - n6 + \sum PUFA - n3}$$

$$IT = \frac{C14:0 + C16:0}{0.5 \times MUFA + 0.5 \times PUFA - n6 + 3 \times PUFA - n3 + \frac{PUFA-n3}{PUFA-n6}}$$

where *MUFA* and *PUFA* correspond to relative concentrations of the monounsaturated fatty acid (*MUFA*) and *PUFA*. The *IA* is related to the plaque formation and to the decrease in the levels of esterified fatty acid, cholesterol, and phospholipids, thereby preventing the appearance of micro- and macro-coronary diseases [33]. The *IT* is related to the tendency to form clots in the blood vessels [33].

For neutral lipids' separation pellets were boiled in water for 5 min to inactivate lipolytic enzymes. The extraction of lipophilic compounds was performed using a mixture of chloroform/methanol/water (1:1:1, *v/v/v*), as previously described [34]. Neutral lipid separation was achieved by thin layer chromatography (TLC) on silica plates (G-60, Merck, VWR) using as solvent a mixture of petroleum ether/ethyl ether/acetic acid (70/30/0.4, *v/v/v*) [35]. Lipids bands were visualized with 0.01% primuline in 80% acetone (*v/v*) under UV light, and the lipid band correspondent to TAG scraped off and transesterified as above-mentioned for total fatty acids, using the scrapped portion as sample [36].

2.7. Statistical Analysis

As normality and homogeneity of variances of our data were not given, pairwise comparisons between different sample groups were performed through nonparametric Kruskal–Wallis tests. These were computed with Statistica software (StataSoft, version 12.5.192.7). Statistical significance was considered at the $p < 0.05$. A multivariate approach was employed to test for variations in the complete photochemical, fatty acid and TAG profiles [32,37,38]. Canonical analysis of principle (CAP) coordinates, using Euclidean distances, were preformed to plot the dissimilarities in a canonical space regarding fatty acids and photochemical studied variables while preforming a cross-validation step and determining the allocation efficiency into the different treatment groups. This multivariate methodology is unaffected by heterogeneous data and frequently used to compare different sample assemblies using the inherent features of each assembly (metabolic traits) [25,32,37,39]. Multivariate statistical analyses were performed using Primer 6 software (version 6.1.13, Plymouth, UK) [40].

3. Results

3.1. Diatom Cell Growth

It was found that RGB and Blue LED treatments resulted in similar cell densities during the growth (Figure 1). However, Red LED treatment caused faster growth rates during the first 48 h ($p < 0.05$), after which the growth rates converged with other treatments.

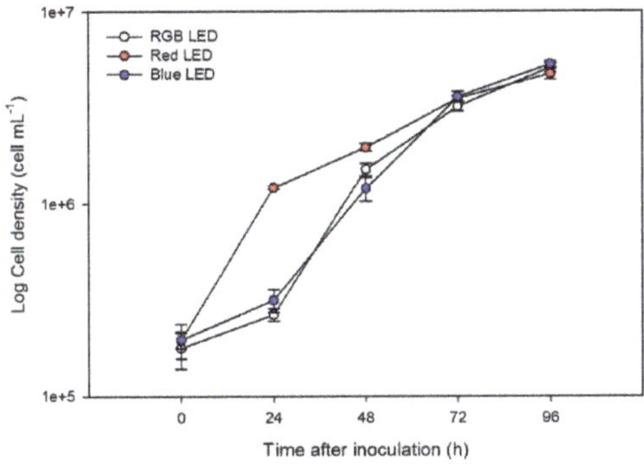

Figure 1. Growth curve of the *P. tricornutum* cultures ($N = 5$) subjected to the three light treatments during the 96 h exposure (average ± standard error).

3.2. Photochemistry at Photosystem II

Rapid light curve measurements (Figure 2a–c) showed that Red LED treatment resulted in reduced relative electron transport rate (rETR) (Figure 2a) when exposed to irradiances above 100 μmol photons m^{-2} s^{-1}, compared to the cultures subjected to the RGB and Blue LED treatments. This resulted in a lower photosynthetic efficiency (Figure 2b) and maximum electron transport rate (Figure 2c) under Red LED treatment. When these parameters are normalized with the pigment weighted light absorption (Q_{phar}), cultures grown under Red LED treatment presented significantly higher Q_{phar} photosynthetic efficiency (Figure 2d). On the other hand, the Q_{phar} maximum ETR of the cells grown under Red LED treatment showed higher values of this parameters as compared to cells exposed to RGB LED treatment (Figure 2e). The cultures grown under monochromatic blue LED illumination showed values of Q_{phar} maximum ETR higher than all the remaining treatments.

Kautsky curve analysis (Figure 3a) revealed differences as a progressive decrease in the overall fluorescence from the cultures exposed to RGB LED treatment, Blue LED treatment and the lowest fluorescence values recorded at the cells grown under only red light. These results in differences at the phenomological energy fluxes (Figure 3b). Both monochromatic light treatments resulted in lower values of absorbed (ABS/CS), trapped (TR/CS) and transported (ET/CS) energy fluxes and also lower values of oxidized PSII reaction centers (RC/CS) when compared to the cultures grown under full spectra illumination (RGB LED). Regarding the dissipated energy flux, this was found to be lower in the cultures grown under monochromatic Red LED illumination. If we analyze the number of active oxygen evolving complexes (AOECs, Figure 3c) located at the donor side of the PSII, it is possible to observe that cultures grown under Red LED illumination resulted in an increase number, while the cells grown under monochromatic Blue LED showed a reduction, both when compared to the cultures grown under full spectra illumination (RGB LED).

Fluorescence profiles from the Kautsky curves in a canonical multivariate approach described the differences between treatments (Figure 4). The CAP analysis showed a clear separation of the samples exposed to the different light treatments (100% classification efficiency), having as basis its fluorescence profile, supporting the differences abovementioned in terms of Kautsky curves-derived parameters.

3.3. Pigment Profiles

The cultures grown under Red LED illumination showed higher contents in chlorophyll *a* and *c*, Fx, DD and DT (Figure 5a). In comparison with the RGB LED grown cultures, the cells cultured under Blue LED also showed higher chlorophylls *a* and *c*, DD and DT concentrations. These differences resulted in higher contents of both total chlorophyll (Figure 5b) and total carotenoids (Figure 5c) in the cultures exposed to the Red and Blue LED treatments.

Applying these pigment concentrations as a whole pigment profile in a multivariate canonical analysis (Figure 6), it is possible to again efficiently distinguish (93.3% classification efficiency) the sample groups, grown under different light treatments, highlighting the differences observed at the light-harvesting and photoprotective pigment level.

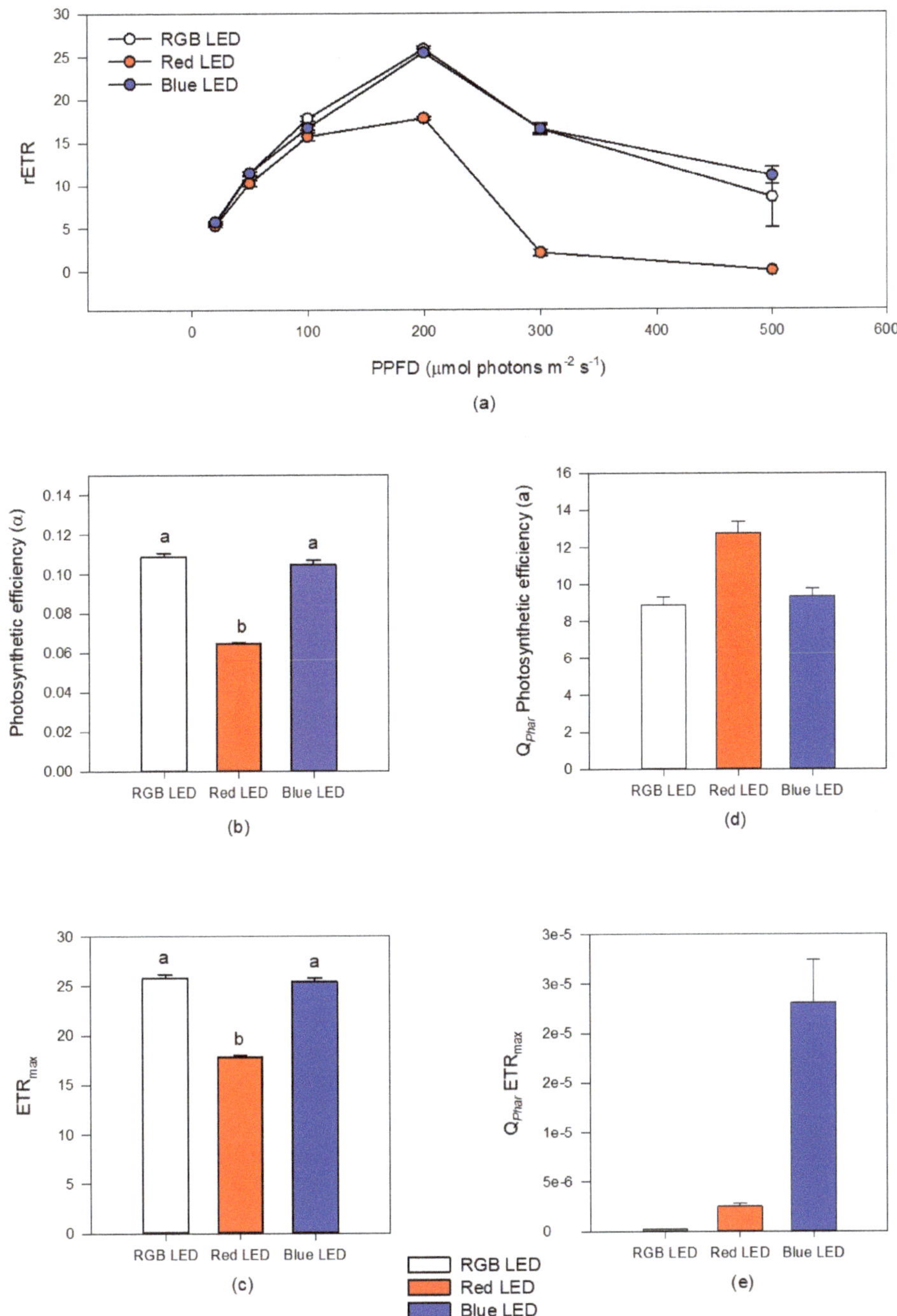

Figure 2. Rapid light curves (**a**), photosynthetic efficiency (**b**) and maximum electron transport rate ((**c**), ETR_{max}) based on incident photosynthetic photon flux density (PPFD) and in the pigment weighted light absorption ((**d**), Q_{Phar} photosynthetic efficiency; (**e**), maximum electron transport rate $Q_{Phar}ETR_{max}$) in *P. tricornutum* cultures ($N = 5$) subjected to the three light treatments at the end of the 96 h exposure period (average $\pm$ standard error, letters denote significant differences between light treatments at $p < 0.05$).

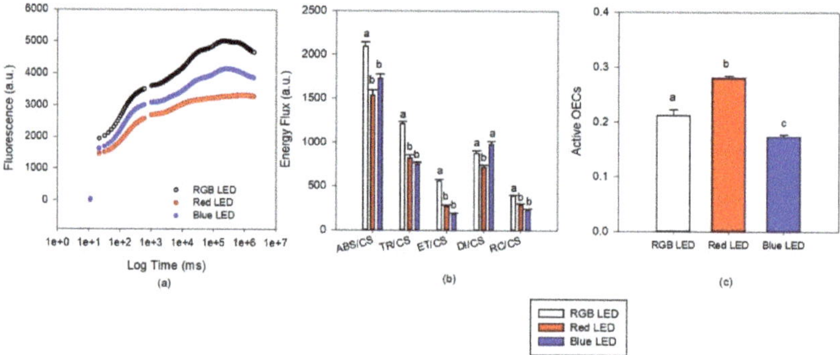

Figure 3. Kautsky transient light curves (**a**), phenomological energy fluxes ((**b**), absorbed (ABS/CS), trapped (TR/CS), transported (ET/CS) and dissipated (DI/CS) energy fluxes and oxidized reaction centers (RC/CS) and active oxygen evolving complexes (**c**) in the *P. tricornutum* cultures (N = 5) subjected to the three light treatments at the end of the 96 h exposure period (average ± standard error, letters denote significant differences between light treatments at $p < 0.05$).

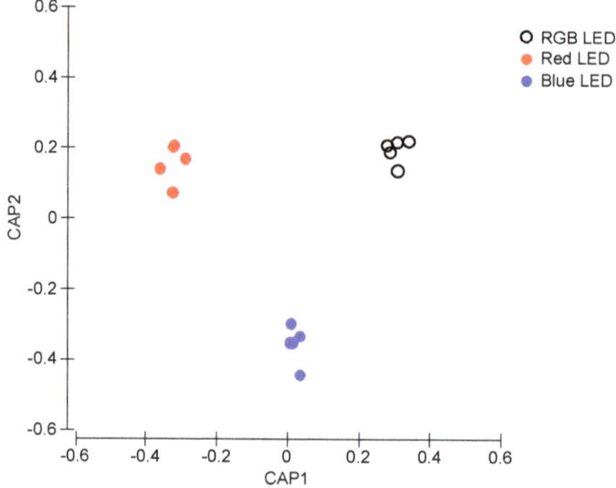

Figure 4. Canonical Analysis of Principal (CAP) components of the *P. tricornutum* cultures ($N = 5$) subjected to the three light treatments, having as basis the full Kaustky fluorescence profile at the end of the 96 h exposure period (100% classification efficiency).

3.4. Total Fatty Acid and TAG Profiles

The culture of diatoms under Blue LED light resulted in cells with a higher total fatty acid content (Figure 7a). Regarding cells composition in terms of individual fatty acid relative abundance (Figure 7b), the diatom cells grown under Red LED illumination, when compared with the cells cultivated under full light spectrum, showed lower contents of 16:0 (palmitic acid), 16:1 (palmitoleic acid) and 18:4 (stearidonic acid, SDA) fatty acids and higher contents in 16:3 (hexadecatrienoic acid) and 20:5 (eicosapentaenoic acid, EPA). Blue monochromatic LED illumination promoted the production of 16:1 and a decrease of EPA cell content, in comparison with the cultures grown under RGB LED illumination. This fatty acid remodeling resulted in significantly lower saturated fatty acid (SFA) and *MUFA* contents in the cells grown under monochromatic Red LED illumination, when compared to the cells grown under full spectrum illumination (Figure 7d). Comparing also with the RGB LED exposed cultures, these same cultures presented significantly higher *PUFA* and unsaturated fatty acid (UFA) contents. Regarding the cells cultivated under Blue LED illumination, their *PUFA* cellular content showed a significant decrease. These alterations combined resulted in differences in *IA* and *IT* (Figure 7c).

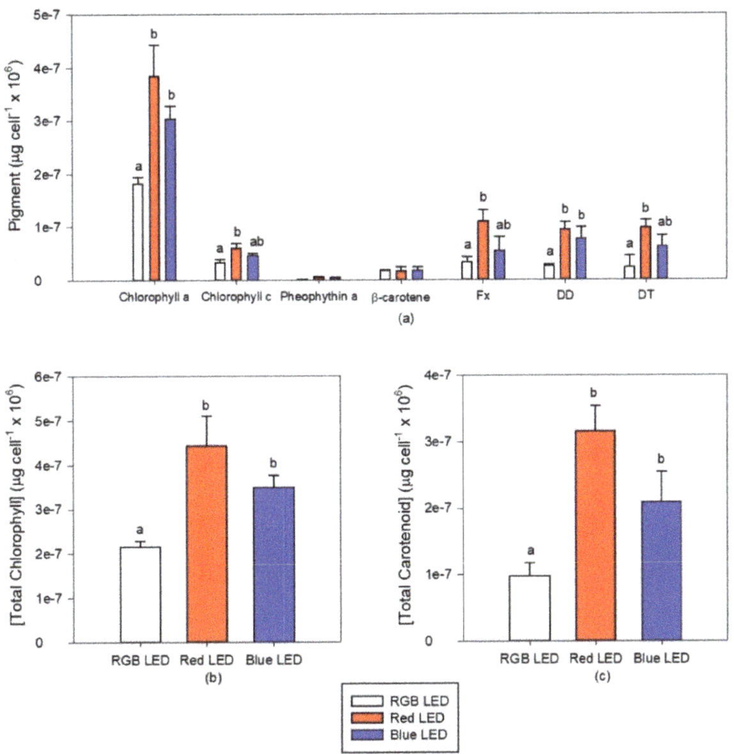

Figure 5. Pigments profiles (**a**), total chlorophyll (**b**) and total carotenoids (**c**) contents of the *P. tricornutum* cultures ($N = 5$) subjected to the three light treatments at the end of the 96 h exposure period (average ± standard error, letters denote significant differences between light treatments at $p < 0.05$).

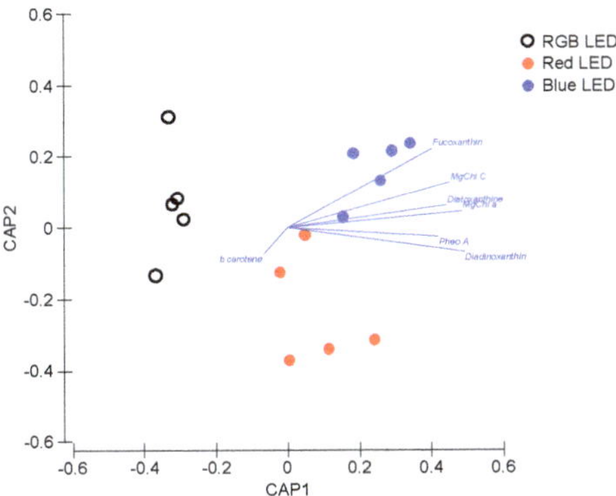

Figure 6. Canonical Analysis of Principal (CAP) components of the *P. tricornutum* cultures ($N = 5$) subjected to the three light treatments, having as basis the full pigment profile at the end of the 96 h exposure period (93.3% classification efficiency).

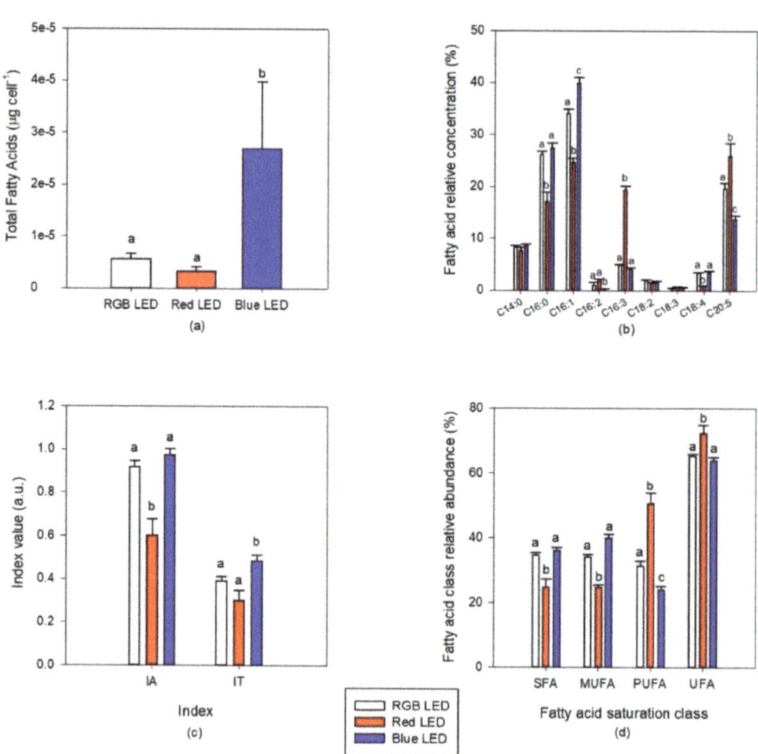

Figure 7. Total fatty acid content (**a**) and profile (**b**), indexes of atherogenicity (*IA*) and thrombogenicity (*IT*) (**c**) and fatty acid unsaturation classes (**d**) of the *P. tricornutum* cultures (*N* = 5) subjected to the three considered light treatments at the end of the 96 h exposure period (average ± standard error, letters denote significant differences between light treatments at $p < 0.05$).

Regarding the atherogenicity capacity of the cultures in preventing the appearance of micro- and macro-coronary diseases, it was found that the cells grown under monochromatic Red LED illumination are favored in this regard showing a significantly lower *IA* value, when compared with both RGB and Blue LED exposed cells. On the other hand, blue light exposure seems to favor the thrombogenicity ability of the culture's fatty acid profile, presenting a higher *IT* value.

As previously computed for the photochemical and pigment data, the fatty acid profiles of the samples were also analyzed in a multivariate canonical approach in order to evaluate if these fatty acid traits are efficient descriptors of the cells culture light conditions (Figure 8). Once again, a clear separation of the samples grown under different light qualities is observed, with the overall canonical analysis presenting a classification efficiency of the samples of 93.3%, having as basis the cells fatty acid profiles.

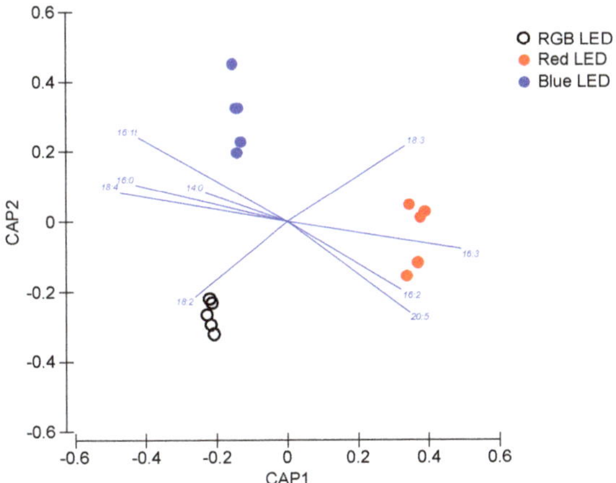

Figure 8. Canonical Analysis of Principal (CAP) components of the *P. tricornutum* cultures (N = 5) subjected to the three light treatments, having as basis the total fatty acid profile (relative concentrations) at the end of the 96 h exposure period (93.3% classification efficiency).

Regarding the TAG fatty acid profile, the different light qualities to which the cultures were subjected also induced some significant differences (Figure 9a). The TAG fatty acid profile of the cultures grown under monochromatic Red LED illumination showed significantly lower values of 14:0 (myristic acid), 16:0, and 16:1 fatty acid relative concentration and significantly higher concentrations of 16:3 (hexadecadienoic acid) and EPA fatty acids. Cultures grown under Blue LED illumination on the other hand showed a significant increase in the TAG concentration of 18:2 (linoleic acid), 18:3 (γ-linolenic acid, GLA) and 18:4. This had inevitable results in the saturation classes of the fatty acid profile of the TAG of the cells cultured at different light conditions (Figure 9b). Significant differences could only be observed in the cells cultured under Red LED illumination, with higher concentration of *PUFA* and lower contents of TAG SFA and *MUFA* in TAG, when compared to the cells cultured under RGB LED illumination. Regarding the TAG cellular content (Figure 9c), this was found to be significantly reduced by the cultivation of the diatom cells under Red LED illumination. In addition, considering the changes above reported regarding the total fatty acids content, the TAG relative content when compared to the total fatty acid in the cells cultivated under red and blue LED illumination decreased to 0.7% and 2.6% respectively versus 5.2% observed in the cells cultured under RGB LED illumination.

Nevertheless, the slight differences observed, the TAG fatty acid profile also proved to efficiently discriminate (100% classification accuracy) the cultures cultivated under different light qualities, indicating that this TAG fatty acid profile is specific of the light used for diatom cultivation and thus can be modified by the illumination wavelength range (Figure 10).

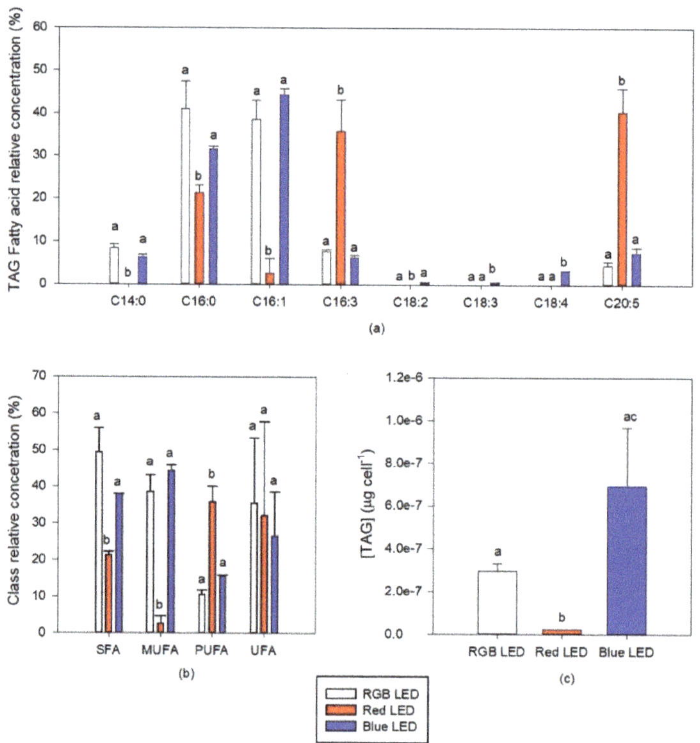

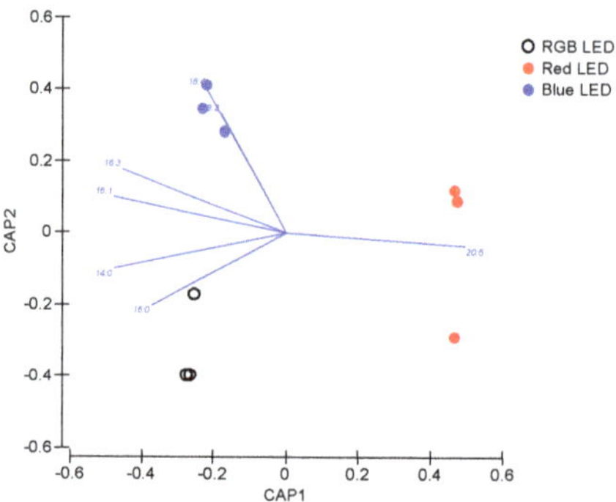

Figure 9. TAG fatty acid profile (**a**) saturation classes (**b**) and cellular content (**c**) of the *P. tricornutum* cultures ($N = 3$) subjected to the three light treatments at the end of the 96 h exposure period (average ± standard error, letters denote significant differences between light treatments at $p < 0.05$).

Figure 10. Canonical Analysis of Principal (CAP) components of the *P. tricornutum* cultures ($N = 5$) subjected to the three light treatments, having as basis the TAG fatty acid profile (relative concentrations) at the end of the 96 h exposure period (100% classification efficiency).

4. Discussion

Diatoms have been suggested as potential high value molecule suppliers for nutraceutical, fuel and aquaculture sectors, by providing bioenergy, food and feed and as supporter of wastewater bioremediation [3]. They are fast growing organisms and require low cultivation conditions while offering high process turnover and economically viable processing alternatives [2]. The current study reports that pigment and fatty acid profiles in the model diatom *Phaeodactylum tricornutum* can be modulated based on different spectral light regimes supplied through LEDs without limiting cell division rates.

At this level, the application of RGB LED illumination versus red and Blue LED illuminations did not lead to different cell densities at the end of the 96 h culture period. Such similarity in cell density at the end of the culture trials might be due to the nutritional limitation in the late phase of the culture, regarded to the high division rates common in the species *P. tricornutum* [41]. However, cultures exposed to red LED illumination presented a faster growth during the first 48 h, indicating a potential role of the light conditions for exponential phase cell growth enhancement. As all light treatments were adjusted to the same total photon flux rates, increased growth during the first days of culture may therefore indicate improved light harvesting under Red LED illumination. Previous studies showed an enhancement of microalgae cell growth cultured under blue light illumination [42] due to higher photosynthetic electron transport, while blue light may also induce higher nonphotochemical quenching [43]. Although most diatom pigments have higher absorption rates for shorter wavelength of light (blue light), at this level no higher growth rates were observed, indicating that given photon flux densities were sufficient to obtain maximum cell density growth. Excessive energy might then be dissipated along the photosynthetic transport chain, limiting accumulation of biomass [44]. When photosynthetic efficiencies are compared to RGB illumination, no apparent differences occur at the end of the experiment. However, when cell growth is compared based on pigment weighted light absorption, some differences become obvious. The cells grown under red LED illumination have higher Q_{phar} photosynthetic efficiency concomitant with the higher growth rates as observed during the first 48 h. This is due to the differences in the light absorption of the different pigment profiles originating from the different light quality exposure. This might be due to the high chlorophyll *a* content, a pigment with a high absorption coefficient at 627 nm (equals red LED dominant wavelength) [30]. In terms of phenomenological energy fluxes the cells grown under red LED illumination showed a reduction in all energy fluxes in a proportional manner, indicating that there is a higher efficiency in the incident light-use. This was in fact previously reported [17], where *P. tricornutum* cells grow under red light showed lower nonphotochemical quenching (lower energy dissipation) and thus higher light-use efficiencies per incident photon. This is also in accordance with the higher dissipated energy flux rates (concomitant with higher nonphotochemical quenching, as reported in [17]) observed in the cells grown under Blue LED illumination. A possible explanation for the observed similar cell densities at the end of the trials under different LED regimes was also reported in the centric diatom *Coscinodiscus granii*, pointing towards illumination conditions supporting maximum diatom cell growth [45]. In sum, in photochemical terms, cells are not substantially affected by the different illumination conditions at the end of the experiment, under the given nutrient conditions and PPFD availabilities. However, it is important to point out that, according to our data, a culture medium may replenish with nutrients at the end of 48 h, as indicated by higher exponential growth under red illumination in the beginning of the trial.

In terms of fatty acid and TAG accumulation, cell cultures under Blue LED light exhibited high total fatty acid and TAG concentrations. This high accumulation of lipids was already reported before, in the diatom *Skeletonema marinoi* [46], pointing out an effective energy and carbon storage strategy [47]. The increase in storage lipids such as TAG, under high frequency blue light has also been attributed to their role in preventing photo-oxidative damages [48]. In fact, and as abovementioned, the cultures subjected to Blue LED illumination showed higher dissipated energy fluxes, also indicative of the need to dissipate excessive energy, another strategy to prevent photo-inhibition.

In terms of bioenergy production, the results also have a relevant role. High amounts of SFAs provide a superior oxidative stability, while PUFAs provide better cold-flow properties at the cost of oxidative stability [49]. Examining the saturation degree of the different cultures, two important characteristics were evident. Both in terms of whole fatty acid and TAG contents, the cells grown under red LED illumination presented a high content in *PUFA*, i.e., higher cold-flow properties. These cultures also showed another important feature in terms of bioenergy production. The fatty acid saturation degree was linearly correlated to the cetane number (CN), a fuel quality parameter which is related to ignition delay and combustion quality of the fuels [49]. At this level also the cells grown under red light illumination showed a substantially higher UFAs percentage, reinforcing the promising characteristics of the fatty acids produced under these experimental conditions. Cultures subjected to blue light illumination presented a higher fatty acid content and in terms of quality for bioenergy applications presented higher saturation degree and thus better characteristics for biodiesel production.

Some nutritional characteristics can also be highlighted for the fatty acid profiles produced under the different light qualities. As abovementioned two indices can be used to evaluate the potential role of certain fatty acid profiles for human health: the *IA* value is related to the plaque formation and to the decrease in the levels of esterified fatty acid, cholesterol, and phospholipids, thereby preventing the appearance of micro- and macro-coronary diseases [33]; the *IT* value is related to the tendency to form clots in the blood vessels [33]. Since saturated FAs are considered to be proatherogenic and UFAs antiatherogenic, a low *IA* ratio is recommended [33]. Regarding *IT*, this index indicates the propensity to form masses in the blood vessels and is defined as the relationship between the prothrombogenic (saturated fatty acids) and the antithrombogenic *MUFA*, n-3 and n-6 *PUFA*). Therefore, a low *IT* value is also desirable [33]. Tuna, a fish considered to have a high nutritional value in terms of its fatty acid profile, has an *IA* and *IT* of approximately 0.7 and 0.3, respectively [33], and is therefore used as a reference. At this level, the cultures grown under red wavelength illumination present a better nutritional quality, with very low *IA* and *IT* values. Moreover, red illumination promoted the accumulation of PUFAs, namely EPA. This fatty acid intake has significant improvements in terms of human health, improving the vascular and neural health [50,51]. Moreover, considering that a large fraction of the developed countries population presents a very low intake of EPA and that the largest source of this *PUFA* is fish intake, another aspect that is highly reduced in western populations, this EPA content acquires a reinforced role as a possible alternative source of these key fatty acids [50,51]. Biosynthesis of EPA is thought to occur in the endoplasmic reticulum via two complementary pathways [52]. In the ω-3 pathway α-Linolenic acid (18:3, ALA) is desaturated to 18:4 which suffers subsequent elongation to 20:4, while in the ω-6 pathway 18:2 is desaturated to GLA and subsequently elongated to eicosatrienoic acid (20:3, ETA) and desaturated to 20:4. A Δ5-desaturase is reported to act in both pathways, converting 20:3 to 20:4 in the ω-6 pathway, and 20:4 to EPA in the ω-3 pathway [52,53]. Overall, cultures grown under red light revealed increased abundance of 16:3 and EPA, a trend which was also reflected in the TAG composition. It is possible that red light induced recycling of plastidal membrane fatty acids through phospholipase A activity given the increased abundance of 16:3 in TAG [54]. The overall increase in C16 fatty acids, namely 16:2 and 16:3, could suggest variations in lipid classes such as monogalactosyldiacylglycerol (MGDG) and digalactosyldiacylglycerol (DGDG), responsible for fucoxanthin chlorophyll proteins (FCP) stabilization and electron transport [55]. The decreased abundance of 18:4 in cultures grown under red illumination could also suggest a prominence of Δ6-elongase activity or gene expression in the ω-3 pathway converting 18:4 to 20:4, rather than GLA to ETA in the ω-6 pathway, and increased Δ5-desaturase activity, resulting in the observed EPA increase. Blue light induced an increase of 16:1 and decrease in EPA content overall; however, the major changes in TAG composition were observed in C18 fatty acids. In sum, cultures grown under red LED illumination presented a high concentration of chloroplastidal fatty acids and long chain PUFAs, while blue light-growth induced the accumulation of EPA precursors in TAG (Figure 11). These results could imply an interaction between light quality, photoreceptors and regulation of genes involved in the lipid and fatty acid metabolism. It is interesting to notice that stearidonic acid, a precursor of EPA with known significant health promoting effects [56], also had its concentration increased in the cultures grown under blue light illumination. In sum, cultures grown under red LED illumination presented a high concentration of long-chain fatty acids, while blue light-growth induced the accumulation of its precursors (Figure 11).

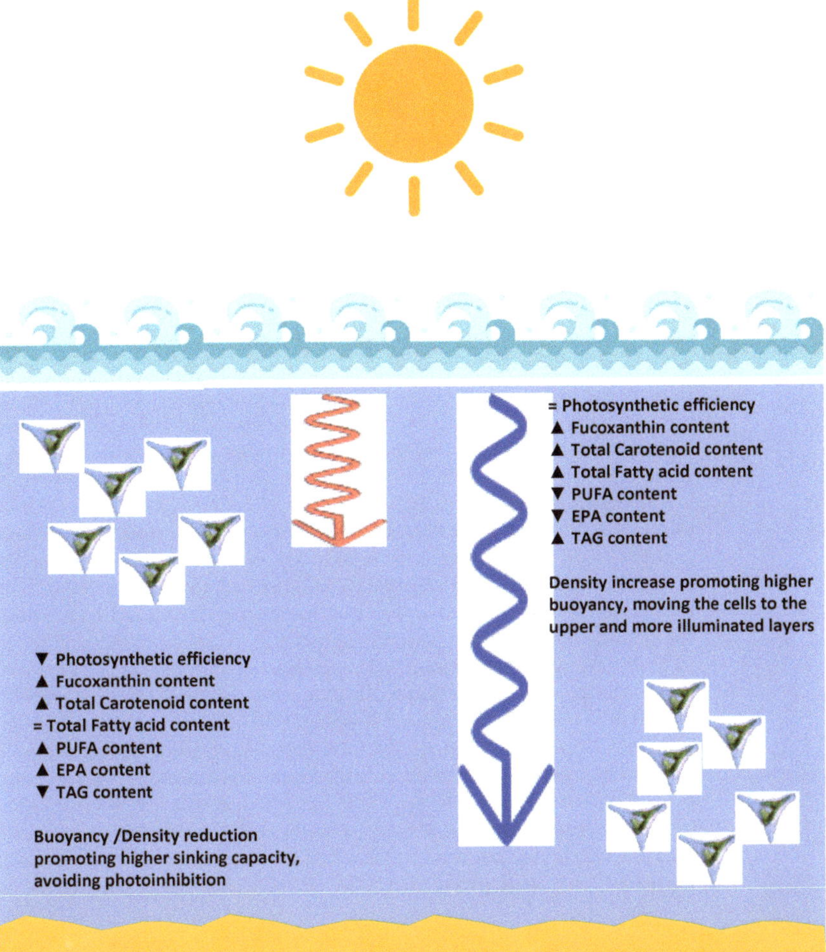

Figure 11. Schematic model of the metabolic changes induced by different wavelengths reaching the different depth layers of the ocean water column (blue and red arrows represent the penetration capacity of the red and blue wavelengths, respectively).

Another group of interesting value-added products produced by marine diatoms are the carotenoids with relevant applications for biotechnology and human health as nutraceuticals [57,58]. From the wide array of pigments produced by diatoms some like β-carotene and Fx have been widely investigated and targeted in studies aiming to enhance its production in microalgae cultures [2,57,58]. Generally speaking, carotenoid pigment production is related to local environmental conditions (such as light conditions and nutrient availability). Fucoxanthin, chlorophyll *a* and *c* are the main light-harvesting pigments in diatoms, whereas β-carotene, DD and DT are mainly involved in photoprotection mechanisms [45]. Regarding the potential nutritional aspects of carotenoids produced these also highlight the potential application of red and blue illumination for improving *P. tricornutum* nutraceutical value. β-carotene is typically the most abundant carotenoid in plants, presenting a recognized antioxidant value, associated with a reduced risk of several diseases including cardiovascular disease [59]. This comes largely from the ability of this pigment to act as a reactive oxygen species (ROS) scavenger [59]. The increase in the pool of β-carotene, precursor of all the other carotenoids, also allows the increased production of other carotenoids such as Fx. Fucoxanthin is a pigment mainly found in brown seaweed and Bacillariophyta (diatoms) [60], and thus any reinforcement in its production is of value-added for industry purposes. This carotenoid has known antioxidant, antiobesity, antidiabetic and anticancer activities [60]. At this level Red LED illumination had a higher effect in the per cell content of this carotenoid, but yet both Blue and Red LED treatments are able to improve diatom Fx production. However, based on the data we conclude that both LED treatments are equally efficient in inducing carotenoid production.

Some similarities observed from the pigment profiles under red and blue light exposed cultures can result from diatom adaptation to the marine environment at different water column depths and are consistent with other studies on phytoplankton [61,62] (Figure 11). In the marine environment, under oligotrophic conditions, light in the blue-green and red spectrum are dominant at higher and lower depths, respectively [63]. Longer wavelengths of the visible light spectrum commonly attenuate faster in water. In consequence, proportions of the red spectral range of light might be higher at subsurface layers where light intensities are higher. Based on the results presented in this study, we propose that diatoms at subsurface layers perceive higher proportions of blue light, inducing less unsaturated TAG remobilization to increase their density and buoyancy. In contrast, diatoms at surface layers might perceive high light flux rates, potentially causing oxidative stress and photodamage. In consequence, carotenoid based photoprotective mechanisms are induced, and the photosynthetic activity is reduced, while cellular energy is dissipated through the consumption of lipid reserves (TAG) leading to reduced cell density and buoyancy reduction. We therefore propose that the proportion of red to blue light are sensed by the diatom cell, either via photoreceptors and/or by physiological feedback loops of the light absorbing and photosynthetic apparatuses (Figure 11) [64]. However, given the complex and diverse light environments in aquatic habitats [15], this proposed mechanisms is speculative and requires further investigation and validation in other planktonic diatom species. Deeper investigation into the natural conditions of light availability in diatom inhabited environments, both in intensity and spectral light composition, may result in better understanding of the apparent complex interplay of cellular light absorption, light perception, photosynthetic activity and other cellular adaptation mechanisms. We propose that such mechanisms may be hijacked for an environmentally friendly production of particular high value products by diatoms, to foster a more sustainable economy in future.

Author Contributions: Conceptualization, J.W.G. and B.D.; investigation, B.D and E.F.; resources, A.R.M.; writing—original draft preparation, B.D.; writing—review and editing, J.W.G., A.R.M., I.C. and E.F.; supervision, A.R.M. and I.C. All authors have read and agreed to the published version of the manuscript.

Funding: The authors would like to thank Fundação para a Ciência e a Tecnologia (FCT) for funding the research via project grants PTDC/CTA-AMB/30056/2017 (OPTOX), UIDB/04292/2020, UIDB/04046/2020. B. Duarte was supported by an investigation contract (CEECIND/00511/2017).

Institutional Review Board Statement: Not applicable.

Informed Consent Statement: Not applicable.

Data Availability Statement: The data presented in this study are available on request from the corresponding author. The data are not publicly available due to being part of an ongoing project.

Conflicts of Interest: The authors declare no conflict of interest.

References

1. Falkowski, P.G.; Laws, E.A.; Barber, R.T.; Murray, J.W. Phytoplankton and Their Role in Primary, New, and Export Production. In *Ocean Biogeochemistry*; Fasham, M.J.R., Ed.; Springer: Berlin/Heidelberg, Germany, 2003; pp. 99–121. ISBN 978-3-642-55844-3.
2. Butler, T.; Kapoore, R.V.; Vaidyanathan, S. *Phaeodactylum tricornutum*: A Diatom Cell Factory. *Trends Biotechnol.* **2020**, *38*, 606–622. [CrossRef]
3. Richmond, A.; Hu, Q. *Handbook of Microalgal Culture*; Richmond, A., Hu, Q., Eds.; John Wiley & Sons, Ltd: Oxford, UK, 2013; ISBN 9781118567166.
4. Kroth, P. Molecular Biology and the Biotechnological Potential of Diatoms. In *Transgenic Microalgae as Green Cell Factories*; León, R., Galván, A., Fernández, E., Eds.; Springer: New York, NY, USA, 2007; pp. 23–33. ISBN 978-0-387-75532-8.
5. Lima, S.; Schulze, P.S.C.; Schüler, L.M.; Rautenberger, R.; Morales-Sánchez, D.; Santos, T.F.; Pereira, H.; Varela, J.C.S.; Scargiali, F.; Wijffels, R.H.; et al. Flashing light emitting diodes (LEDs) induce proteins, polyunsaturated fatty acids and pigments in three microalgae. *J. Biotechnol.* **2021**, *325*, 15–24. [CrossRef]
6. Vílchez, C.; Forján, E.; Cuaresma, M.; Bédmar, F.; Garbayo, I.; Vega, J.M. Marine Carotenoids: Biological Functions and Commercial Applications. *Mar. Drugs* **2011**, *9*, 319–333. [CrossRef] [PubMed]
7. Fu, W.; Wichuk, K.; Brynjólfsson, S. Developing diatoms for value-added products: Challenges and opportunities. *New Biotechnol.* **2015**, *32*, 547–551. [CrossRef]
8. Xiang, S.; Liu, F.; Lin, J.; Chen, H.; Huang, C.; Chen, L.; Zhou, Y.; Ye, L.; Zhang, K.; Jin, J.; et al. Fucoxanthin Inhibits β-Amyloid Assembly and Attenuates β-Amyloid Oligomer-Induced Cognitive Impairments. *J. Agric. Food Chem.* **2017**, *65*, 4092–4102. [CrossRef] [PubMed]
9. Peng, J.; Yuan, J.-P.; Wu, C.-F.; Wang, J.-H. Fucoxanthin, a marine carotenoid present in brown seaweeds and diatoms: Metabolism and bioactivities relevant to human health. *Mar. Drugs* **2011**, *9*, 1806–1828. [CrossRef] [PubMed]
10. Guschina, I.A.; Harwood, J.L. Lipids and lipid metabolism in eukaryotic algae. *Prog. Lipid Res.* **2006**, *45*, 160–186. [CrossRef] [PubMed]
11. Shahidi, F.; Ambigaipalan, P. Omega-3 Polyunsaturated Fatty Acids and Their Health Benefits. *Annu. Rev. Food Sci. Technol.* **2018**, *9*, 345–381. [CrossRef]
12. Hildebrand, M.; Davis, A.K.; Smith, S.R.; Traller, J.C.; Abbriano, R. The place of diatoms in the biofuels industry. *Biofuels* **2012**, *3*, 221–240. [CrossRef]
13. Blanken, W.; Cuaresma, M.; Wijffels, R.H.; Janssen, M. Cultivation of microalgae on artificial light comes at a cost. *Algal Res.* **2013**, *2*, 333–340. [CrossRef]
14. Stomp, M.; Huisman, J.; Stal, L.J.; Matthijs, H.C.P. Colorful niches of phototrophic microorganisms shaped by vibrations of the water molecule. *ISME J.* **2007**, *1*, 271–282. [CrossRef] [PubMed]
15. Kirk, J.T.O. Light and Photosynthesis in Aquatic Ecosystems, 3rd ed.Cambridge University Press: Cambridge, UK, 2010; ISBN 9780521151757.
16. Shikata, T.; Nukata, A.; Yoshikawa, S.; Matsubara, T.; Yamasaki, Y.; Shimasaki, Y.; Oshima, Y.; Honjo, T. Effects of light quality on initiation and development of meroplanktonic diatom blooms in a eutrophic shallow sea. *Mar. Biol.* **2009**, *156*, 875–889. [CrossRef]
17. Jungandreas, A.; Costa, B.S.; Jakob, T.; von Bergen, M.; Baumann, S.; Wilhelm, C. The Acclimation of *Phaeodactylum tricornutum* to Blue and Red Light Does Not Influence the Photosynthetic Light Reaction but Strongly Disturbs the Carbon Allocation Pattern. *PLoS ONE* **2014**, *9*, e99727. [CrossRef]
18. Herbstová, M.; Bína, D.; Kaňa, R.; Vácha, F.; Litvín, R. Red-light phenotype in a marine diatom involves a specialized oligomeric red-shifted antenna and altered cell morphology. *Sci. Rep.* **2017**, *7*, 11976. [CrossRef] [PubMed]
19. Guillard, R.R.L.; Ryther, J.H. Studies of Marine Planktonic Diatoms: I. Cyclotella Nana Hustedt, and Detonula Confervacea (Cleve) Gran. *Can. J. Microbiol.* **1962**, *8*, 229–239. [CrossRef] [PubMed]
20. Feijão, E.; Gameiro, C.; Franzitta, M.; Duarte, B.; Caçador, I.; Cabrita, M.T.; Matos, A.R. Heat wave impacts on the model diatom *Phaeodactylum tricornutum*: Searching for photochemical and fatty acid biomarkers of thermal stress. *Ecol. Indic.* **2018**, *95*, 1026–1037. [CrossRef]
21. OECD Guidelines for the Testing of Chemicals. Freshwater Alga and Cyanobacteria, Growth Inhibition Test. *Organ. Econ. Coop. Dev.* **2011**, 1–25.
22. Tachi, N.; Hashimoto, Y.; Ogino, N. Vitrectomy for macular edema combined with retinal vein occlusion. *Doc. Ophthalmol.* **1999**, *97*, 465–469. [CrossRef]
23. Srivastava, A.; Strasser, R.J. Survival strategies of plants to cope with the stress of daily atmospheric changes. In *Crop Improvement for Food Security*; Behl, R.K., Punia, M.S., Lather, B.P.S., Eds.; SSARM: Hisar, Idina, 1999; pp. 60–71.
24. Strasser, R.J.; Tsimilli-Michael, M.; Srivastava, A. Analysis of the fluorescence transient. In *Chlorophyll Fluorescence: A Signature of Photosynthesis. Advances in Photosynthesis and Respiration Series*; Papageorgiou, G.C., Govindjee, Eds.; Springer: Dordrecht, The Netherlands, 2004; pp. 321–362.
25. Cabrita, M.T.; Duarte, B.; Gameiro, C.; Godinho, R.M.; Caçador, I. Photochemical features and trace element substituted chlorophylls as early detection biomarkers of metal exposure in the model diatom *Phaeodactylum tricornutum*. *Ecol. Indic.* **2018**, *95*, 1038–1052. [CrossRef]

26. Cabrita, M.T.; Gameiro, C.; Utkin, A.B.; Duarte, B.; Caçador, I.; Cartaxana, P. Photosynthetic pigment laser-induced fluorescence indicators for the detection of changes associated with trace element stress in the diatom model species *Phaeodactylum tricornutum*. *Environ. Monit. Assess.* **2016**, *188*, 285. [CrossRef]

27. Küpper, H.; Seibert, S.; Parameswaran, A. Fast, sensitive, and inexpensive alternative to analytical pigment HPLC: Quantification of chlorophylls and carotenoids in crude extracts by fitting with Gauss peak spectra. *Anal. Chem.* **2007**, *79*, 7611–7627. [CrossRef]

28. Prins, A.; Deleris, P.; Hubas, C.; Jesus, B. Effect of Light Intensity and Light Quality on Diatom Behavioral and Physiological Photoprotection. *Front. Mar. Sci.* **2020**, *7*. [CrossRef]

29. Gilbert, M.; Domin, A.; Becker, A.; Wilhelm, C. Estimation of Primary Productivity by Chlorophyll a in vivo Fluorescence in Freshwater Phytoplankton. *Photosynthetica* **2000**, *38*, 111–126. [CrossRef]

30. Clementson, L.A.; Wojtasiewicz, B. Dataset on the absorption characteristics of extracted phytoplankton pigments. *Data Br.* **2019**, *24*, 103875. [CrossRef] [PubMed]

31. Matos, A.R.; Hourton-Cabassa, C.; Ciçek, D.; Rezé, N.; Arrabaça, J.D.; Zachowski, A.; Moreau, F. Alternative oxidase involvement in cold stress response of *Arabidopsis thaliana* fad2 and FAD3$^+$ cell suspensions altered in membrane lipid composition. *Plant Cell Physiol.* **2007**, *48*, 856–865. [CrossRef] [PubMed]

32. Duarte, B.; Prata, D.; Matos, A.R.; Cabrita, M.T.; Caçador, I.; Marques, J.C.; Cabral, H.N.; Reis-Santos, P.; Fonseca, V.F. Ecotoxicity of the lipid-lowering drug bezafibrate on the bioenergetics and lipid metabolism of the diatom *Phaeodactylum tricornutum*. *Sci. Total Environ.* **2019**, *650*, 2085–2094. [CrossRef]

33. Garaffo, M.A.; Vassallo-Agius, R.; Nengas, Y.; Lembo, E.; Rando, R.; Maisano, R.; Dugo, G.; Giuffrida, D. Fatty Acids Profile, Atherogenic (IA) and Thrombogenic (IT) Health Lipid Indices, of Raw Roe of Blue Fin Tuna (*Thunnus thynnus* L.) and Their Salted Product "Bottarga". *Food Nutr. Sci.* **2011**, *2*, 736–743.

34. Lee, H.Y.; Bahn, S.C.; Kang, Y.-M.M.; Lee, K.H.; Kim, H.J.; Noh, E.K.; Palta, J.P.; Shin, J.S.; Ryu, S.B. Secretory Low Molecular Weight Phospholipase A 2 Plays Important Roles in Cell Elongation and Shoot Gravitropism in Arabidopsis. *Plant Cell* **2003**, *15*, 1990–2002. [CrossRef] [PubMed]

35. Matos, A.R.A.R.; Gigon, A.; Laffray, D.; Pêtres, S.; Zuily-Fodil, Y.; Pham-Thi, A.-T. Effects of progressive drought stress on the expression of patatin-like lipid acyl hydrolase genes in Arabidopsis leaves. *Physiol. Plant.* **2008**, *134*, 110–120. [CrossRef]

36. Laureano, G.; Figueiredo, J.; Cavaco, A.R.; Duarte, B.; Caçador, I.; Malhó, R.; Silva, M.S.; Matos, A.R.; Figueiredo, A. Author Correction: The interplay between membrane lipids and phospholipase A family members in grapevine resistance against Plasmopara viticola. *Sci. Rep.* **2019**, *9*, 6731. [CrossRef] [PubMed]

37. Duarte, B.; Pedro, S.; Marques, J.C.; Adão, H.; Caçador, I. *Zostera noltii* development probing using chlorophyll a transient analysis (JIP-test) under field conditions: Integrating physiological insights into a photochemical stress index. *Ecol. Indic.* **2017**, *76*, 219–229. [CrossRef]

38. Feijão, E.; de Carvalho, R.; Duarte, I.A.; Matos, A.R.; Cabrita, M.T.; Novais, S.C.; Lemos, M.F.L.; Caçador, I.; Marques, J.C.; Reis-Santos, P.; et al. Fluoxetine Arrests Growth of the Model Diatom *Phaeodactylum tricornutum* by Increasing Oxidative Stress and Altering Energetic and Lipid Metabolism. *Front. Microbiol.* **2020**, *11*, 1803. [CrossRef]

39. Duarte, B.; Cabrita, M.T.; Vidal, T.; Pereira, J.L.; Pacheco, M.; Pereira, P.; Canário, J.; Gonçalves, F.J.M.; Matos, A.R.; Rosa, R.; et al. Phytoplankton community-level bio-optical assessment in a naturally mercury contaminated Antarctic ecosystem (Deception Island). *Mar. Environ. Res.* **2018**, *140*, 412–421. [CrossRef] [PubMed]

40. Clarke, K.R.; Gorley, R.N. PRIMER v6: User Manual/Tutorial. *Prim. Plymouth UK* **2006**, 192.

41. Yang, R.; Wei, D.; Xie, J. Diatoms as cell factories for high-value products: Chrysolaminarin, eicosapentaenoic acid, and fucoxanthin. *Crit. Rev. Biotechnol.* **2020**, *40*, 993–1009. [CrossRef]

42. Sirisuk, P.; Ra, C.-H.H.; Jeong, G.-T.T.; Kim, S.-K.K. Effects of wavelength mixing ratio and photoperiod on microalgal biomass and lipid production in a two-phase culture system using LED illumination. *Bioresour. Technol.* **2018**, *253*, 175–181. [CrossRef] [PubMed]

43. Goessling, J.W.; Cartaxana, P.; Kühl, M. Photo-protection in the centric diatom *Coscinodiscus granii* is not controlled by chloroplast high-light avoidance movement. *Front. Mar. Sci.* **2016**, *2*. [CrossRef]

44. Jeong, H.; Lee, J.; Cha, M. Energy efficient growth control of microalgae using photobiological methods. *Renew. Energy* **2013**, *54*, 161–165. [CrossRef]

45. Su, Y. The effect of different light regimes on pigments in *Coscinodiscus granii*. *Photosynth. Res.* **2019**, *140*, 301–310. [CrossRef]

46. Chandrasekaran, R.; Barra, L.; Carillo, S.; Caruso, T.; Corsaro, M.M.; dal Piaz, F.; Graziani, G.; Corato, F.; Pepe, D.; Manfredonia, A.; et al. Light modulation of biomass and macromolecular composition of the diatom *Skeletonema marinoi*. *J. Biotechnol.* **2014**, *192*, 114–122. [CrossRef]

47. Fábregas, J.; Maseda, A.; Domínguez, A.; Ferreira, M.; Otero, A. Changes in the cell composition of the marine microalga, *Nannochloropsis gaditana*, during a light:dark cycle. *Biotechnol. Lett.* **2002**, *24*, 1699–1703. [CrossRef]

48. Solovchenko, A.E. Physiological role of neutral lipid accumulation in eukaryotic microalgae under stresses. *Russ. J. Plant Physiol.* **2012**, *59*, 167–176. [CrossRef]

49. Ramos, M.J.; Fernández, C.M.; Casas, A.; Rodríguez, L.; Pérez, Á. Influence of fatty acid composition of raw materials on biodiesel properties. *Bioresour. Technol.* **2009**, *100*, 261–268. [CrossRef] [PubMed]

50. Calder, P.C. Functional Roles of Fatty Acids and Their Effects on Human Health. *J. Parenter. Enter. Nutr.* **2015**, *39*, 18S–32S. [CrossRef] [PubMed]

51. Calder, P.C. Very long-chain n -3 fatty acids and human health: Fact, fiction and the future. *Proc. Nutr. Soc.* **2018**, *77*, 52–72. [CrossRef]
52. Mühlroth, A.; Li, K.; Røkke, G.; Winge, P.; Olsen, Y.; Hohmann-Marriott, M.; Vadstein, O.; Bones, A. Pathways of Lipid Metabolism in Marine Algae, Co-Expression Network, Bottlenecks and Candidate Genes for Enhanced Production of EPA and DHA in Species of Chromista. *Mar. Drugs* **2013**, *11*, 4662–4697. [CrossRef]
53. Dolch, L.-J.; Maréchal, E. Inventory of Fatty Acid Desaturases in the Pennate Diatom *Phaeodactylum tricornutum*. *Mar. Drugs* **2015**, *13*, 1317–1339. [CrossRef] [PubMed]
54. Matos, A.R.; Pham-Thi, A.-T. Lipid deacylating enzymes in plants: Old activities, new genes. *Plant Physiol. Biochem.* **2009**, *47*, 491–503. [CrossRef]
55. Feijão, E.; Franzitta, M.; Cabrita, M.T.; Caçador, I.; Duarte, B.; Gameiro, C.; Matos, A.R. Marine heat waves alter gene expression of key enzymes of membrane and storage lipids metabolism in *Phaeodactylum tricornutum*. *Plant Physiol. Biochem.* **2020**, *156*, 357–368. [CrossRef]
56. Walker, C.G.; Jebb, S.A.; Calder, P.C. Stearidonic acid as a supplemental source of ω-3 polyunsaturated fatty acids to enhance status for improved human health. *Nutrition* **2013**, *29*, 363–369. [CrossRef] [PubMed]
57. Koller, M.; Muhr, A.; Braunegg, G. Microalgae as versatile cellular factories for valued products. *Algal Res.* **2014**, *6*, 52–63. [CrossRef]
58. McClure, D.D.; Luiz, A.; Gerber, B.; Barton, G.W.; Kavanagh, J.M. An investigation into the effect of culture conditions on fucoxanthin production using the marine microalgae *Phaeodactylum tricornutum*. *Algal Res.* **2018**, *29*, 41–48. [CrossRef]
59. Helmersson, J.; Ärnlv, J.; Larsson, A.; Basu, S. Low dietary intake of β-carotene, α-tocopherol and ascorbic acid is associated with increased inflammatory and oxidative stress status in a Swedish cohort. *Br. J. Nutr.* **2009**, *101*, 1775–1782. [CrossRef]
60. Miyashita, K.; Beppu, F.; Hosokawa, M.; Liu, X.; Wang, S. Nutraceutical characteristics of the brown seaweed carotenoid fucoxanthin. *Arch. Biochem. Biophys.* **2020**, *686*, 108364. [CrossRef]
61. Humphrey, G.F. The effect of the spectral composition of light on the growth, pigments, and photosynthetic rate of unicellular marine algae. *J. Exp. Mar. Bio. Ecol.* **1983**, *66*, 49–67. [CrossRef]
62. Mouget, J.-L.; Rosa, P.; Tremblin, G. Acclimation of *Haslea ostrearia* to light of different spectral qualities-confirmation of 'chromatic adaptation' in diatoms. *J. Photochem. Photobiol. B Biol.* **2004**, *75*. [CrossRef] [PubMed]
63. Depauw, F.A.; Rogato, A.; d'Alcala, M.R.; Falciatore, A. Exploring the molecular basis of responses to light in marine diatoms. *J. Exp. Bot.* **2012**, *63*, 1575–1591. [CrossRef] [PubMed]
64. Brunet, C.; Chandrasekaran, R.; Barra, L.; Giovagnetti, V.; Corato, F.; Ruban, A.V. Spectral Radiation Dependent Photoprotective Mechanism in the Diatom *Pseudo-nitzschia multistriata*. *PLoS ONE* **2014**, *9*, e87015. [CrossRef] [PubMed]

 *applied sciences*

Article

Application of Box-Behnken Design and Desirability Function for Green Prospection of Bioactive Compounds from *Isochrysis galbana*

Mari Carmen Ruiz-Domínguez [1,*]☉, Pedro Cerezal [1], Francisca Salinas [1], Elena Medina [1] and Gabriel Renato-Castro [2]

[1] Laboratorio de Microencapsulación de Compuestos Bioactivos (LAMICBA, acronym in Spanish), Departamento de Ciencias de los Alimentos y Nutrición, Facultad de Ciencias de la Salud, Universidad de Antofagasta #02800, Antofagasta 1240000, Chile; pedro.cerezal@uantof.cl (P.C.); francisca.salinas@uantof.cl (F.S.); elena.medina.perez@ua.cl (E.M.)

[2] Departamento de Investigación Biotecnológica Empresa Microalgas Oleas de México S.A. de C.V. Parque Tecnológico–ITESO, Guadalajara 45221, Mexico; gabriel.renato@oleopalma.com.mx

* Correspondence: maria.ruiz@uantof.cl; Tel.: +56-552-633-660

Received: 9 March 2020; Accepted: 15 April 2020; Published: 17 April 2020

Abstract: A microalga, *Isochrysis galbana*, was chosen in this study for its potent natural antioxidant composition. A broad bioactive compounds spectrum such as carotenoids, fatty acid polyunsaturated (PUFA), and antioxidant activity are described with numerous functional properties. However, most of the optimization of extraction use toxic solvents or consume a lot of it becoming an environmental concern. In this research, a Box-Behnken design with desirability function was used to prospect the bioactive composition by supercritical fluid extraction (SFE) after performing the kinetics curve to obtain the optimal extraction time minimizing operational costs in the process. The parameters studied were: pressure (20–40 MPa), temperature (40–60 °C), and co-solvent (0–8% ethanol) with a CO_2 flow rate of 7.2 g/min for 120 min. The response variables evaluated in *I. galbana* were extraction yield, carotenoids content and recovery, total phenols, antioxidant activity (TEAC method, trolox equivalents antioxidant capacity method), and fatty acid profile and content. In general, improvement in all variables was observed using an increase in ethanol concentration used as a co-solvent (8% v/v ethanol) high pressure (40 MPa), and moderately high temperature (50 °C). The fatty acids profile was rich in polyunsaturated fatty acid (PUFA) primarily linoleic acid (C18:2) and linolenic acid (C18:3). Therefore, *I. galbana* extracts obtained by supercritical fluid extraction showed relevant functional ingredients for use in food and nutraceutical industries.

Keywords: microalgae; fucoxanthin; fatty acids; antioxidant; supercritical CO_2 extraction; co-solvent.

1. Introduction

Lately, the interest in studying the food applications of microalgae has increased significantly due to high nutritional value and a vast variety of novel metabolites with numerous innovative food applications. These could be considered such as nutraceuticals, food supplement, functional ingredients, or functional foods. Continually, new interpretations, and dynamic situation in the food sector provoke that these applications are situated between ordinary food and medical drugs [1]. In addition, huge interest has emerged among consumers and the nutrition industry in novel products that can promote good health, improve the state of wellbeing, and decrease the risk of diseases [1,2]. Diabetes, cardiovascular diseases, obesity, hypertension, cancer, or depression are some examples of diseases where these novel products combined with healthy lifestyle could modulate them [1,2]. Although microalgae have been used for centuries as a source of nutrition [3], many species such as *Arthrospira maxima*, *A. platensis*, *Chlorella vulgaris*, *Haematococcus pluvialis*, *Isochrysis galbana*, *Scenedesmus* sp., *Porphyridium cruentum*, or *Phaeodactylum tricornutum* have been recently included for their composition [4–6]. Proteins, amino acids, polysaccharides, pigments, carotenoids (β-carotene or astaxantina), vitamins, and fatty acids between other are compounds described for theirs beneficial health effects such as: anti-inflammatory, protection of UV radiation, immune system, arthritis, Alzheimer's disease, or cancer or applications in pharmaceutical or cosmetics industries (natural colorants or anti-aging products) [1,4,5,7].

This current research particularly focused on *Isochrysis galbana* (Phylum: Haptophyta). This marine microalga has served in the aquaculture industry as a feed of bivalves, fish larvae, crustaceans, and mollusks for years [8]. Recently, this microalga has also been of interest in adjusting the composition of significant biomolecules, such as polysaccharides, fatty acid, carotenoids, vitamin, and sterols which are bioactive compounds that elicit positive nutrition in human foods or animal feed. Further, these molecules have demonstrated their therapeutic potential against several diseases like cancer, diabetes, cardiovascular and infectious diseases, among others [9,10]. *I. galbana* has also been described as a significant source of vitamin A and E, folic acid, nicotinic acid, pantothenic acid, biotin, thiamin, riboflavin, pyridoxine, cobalamin, chlorophyll (a and c), fucoxanthin, and diadinoxanthin [11].

I. galbana harbors specific carotenoid fucoxanthin and has high content of bioactive compounds including polyunsaturated fatty acids (PUFA) profile such as docosahexaenoic acid (DHA). Fucoxanthin is also present in brown seaweed or several diatoms, such as *Phaeodactylum tricornutum*, and its activity as an anti-inflammatory, antioxidant, and anticancer has been demonstrated in several studies [12]. Additionally, fucoxanthin is also able to modulate certain genes implicated in the cell metabolism, a property seemingly essential for good health [13,14]. In addition, the presence of PUFA is beneficial because they contribute to the production of prostaglandins, and/or thromboxanes which are biologically active substances that play important roles in the reduction of cholesterol and triglycerides in the blood as well as prevention of cardiovascular diseases, atherosclerosis, skin diseases, and arthritis [15,16]. PUFAs should be included in the daily diet because they cannot be synthesized by humans or animals per se [16].

For the extraction of bioactive compounds, an ideal extraction method should result in a rapid quantitative recovery of the target without degradation, and the removal of the solvent post-extraction should be easy and rapid. Many bioactive natural products are thermolabile and can degrade while using traditional extraction methods. For these reasons, supercritical fluids extraction (SFE) through carbon dioxide has been demonstrated as an effective method for the extraction of bioactive compounds [9,17]. The primary advantages of this green extraction are: (i) the possibility to change the density of the fluid through controlled pressure and/or temperature accompanied by modified extraction solubility, (ii) use of green solvent generally recognized as safe (GRAS), and (iii) lower extraction times with enhanced extraction yield [9,18].

Therefore, the aim of this study was to extract and investigate the total carotenoids (fucoxanthin represented as main total carotenoids by spectrophotometry method) present in *I. galbana* using the supercritical fluid extraction technique through carbon dioxide. Parameters of temperature (40–60 °C), pressure (20–40 MPa), and percentage of co-solvent (0–8% ethanol) were controlled based in other previous studies [9]. The extraction yield, total carotenoids content and its recovery, total phenols, antioxidant activity (TEAC, trolox equivalents antioxidant capacity), and fatty acid profile of the obtained extracts were determined. The kinetic study was also performed to control the optimal extraction time in the prospection of total carotenoids from *I. galbana*. Finally, the optimal parameters are recommended for time and extraction conditions of total carotenoids and other compounds present in *I. galbana*, which have significant potential in functional and biotechnological applications minimizing operational costs in the process, for example, the volume of ethanol.

2. Materials and Methods

2.1. Chemicals and Samples

The microalga *I. galbana* was selected for this research. Mexican Company "Microalgas Oleas de México S.A." (Guadalajara, Mexico) kindly donated the dry biomass for this research. The microalga was cultured under a controlled condition and harvested in its exponential growth phase. The biomass was dried under the freeze-dry system such as Labconco FreeZone 2.5 L Benchtop Dry System (Labconco, Kansas City, MO, USA), packed in vacuum sealing plastic bags, and stored at 4 ± 2 °C in darkness until use. The chemical materials used for supercritical fluid extraction (SFE) were carbon dioxide (99% purity), purchased from Indura Group Air Products (Santiago, Chile) and ethanol (99.5%), from Merck (Darmstadt, Germany). Other chemicals such as ultrapure water, fucoxanthin standard, gallic acid, 6-hydroxy-2,5,7,8-tetramethylchroman-2-carboxylic acid (Trolox, ≥ 97%), 2,2-azino-bis (3-ethylbenzothiazoline-6-sulfonic acid (ABTS ≥ 99%), and Folin–Ciocalteu phenol reagent were purchased from Sigma-Aldrich (Santiago, Chile). Chromatographic grade ethyl acetate, water, acetonitrile, methanol, and n-hexane were purchased from Sigma-Aldrich (Santiago, Chile). For fatty acid identification and quantification, a standard fatty acid methyl ester (FAME) mix, C4-C24, supplied by Supelco Analytical (Bellefonte, PA, USA) was used, and tripentadecanoin > 99% (Nu-Check Pre, Inc., Elysian, MN, USA) was used as the internal standard.

2.2. Supercritical Fluid Extraction

The extractions were carried out using a Speed Helix supercritical extractor (Applied Separation, Allentown, PA, USA) as per the scheme presented in Figure 1. For each extraction, 2 g of freeze-dried biomass of *I. galbana* was used, previously ground and sieved using a standard sieve of 35 mesh of the Tyler series (particle size ≤ 0.354 mm), along with polypropylene wool and glass beads (Φ = 1 mm), which was then inserted into stain-steel extraction cell of 24 mL.

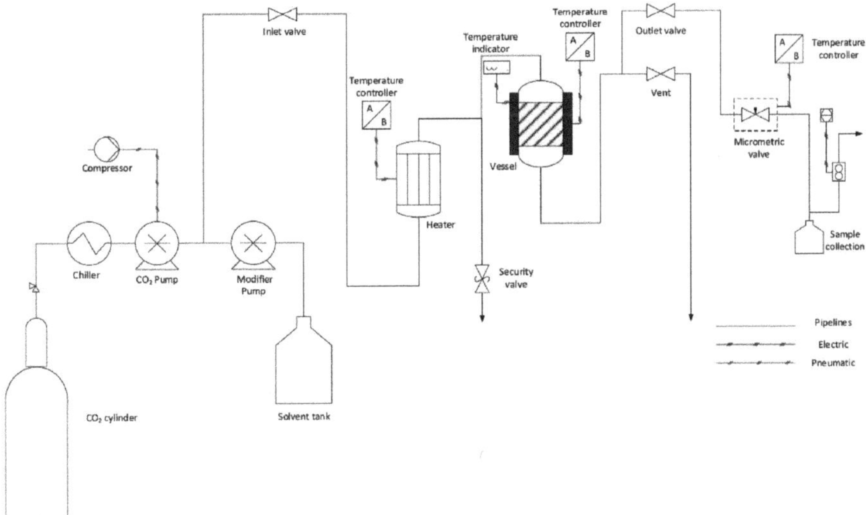

Figure 1. Diagram of the supercritical fluid extraction (SFE) equipment (Applied Separations, Speed, Allentown, PA, USA). The main parts are: CO_2 cylinder; CO_2 pump; compressor; modifier pump; solvent tank; inlet valve; heater; extraction vessel and oven vessel; micrometric valve; sample collection.

In all cases, a flow rate of 7.2 g/min was maintained for CO_2 and each extraction was carried out for 120 min. Extraction conditions of the microalga were selected based on preliminary kinetics assays with *I. galbana* and were set for 150 min to ensure the complete removal of bioactive compounds.

The resulting extracts were collected in vials and the residual ethanol was evaporated under an N_2 gas stream by Flexivap Work Station (Model 109A YH-1, Glas-Col, Terre Haute, IN, USA) for calculating extraction yield. Then, dried extracts were stored at −20 °C and protected from light until further analysis.

2.3. Experimental Design

A Box-Behnken design was implemented in random run order (Table 1), generating 15 experimental conditions tested (Table 2). As per this design, three factors were evaluated at three different experimental levels of temperature (40–60 °C), pressure (20–40 MPa), and percentage of ethanol as a co-solvent (0–8% v/v). The effects of the factors on different responses, including extraction yield (Y), total carotenoids content (TCC) and its recovery (TC recovery), total phenolic content (TPC), antioxidant activity (TEAC assay), and fatty acids profile (FAMEs) were studied. The experimental design and data analysis were carried out using response surface methodology (RSM) with Statgraphics Centurion XVI®(StatPoint Technologies, Inc., Warrenton, VA, USA) software.

In a design that involves three factors X1, X2, and X3, the mathematical relationship of the response with these factors is approximated by the quadratic polynomial equation named second degree Equation (1) described below:

$$Y = \beta_0 + \beta_1 X_1 + \beta_2 X_2 + \beta_3 X_3 + \beta_{12} X_1 X_2 + \beta_{13} X_1 X_3 + \beta_{23} X_2 X_3 + \beta_{11} X_1^2 + \beta_{22} X_2^2 + \beta_{33} X_3^2 \qquad (1)$$

where Y = estimate response; β0 = constant; β1, β2, and β3 = linear coefficients; β12, β13, and β23 = interaction coefficients between the three factors; β11, β22, and β33 = quadratic coefficients. Multiple regression analysis is done to obtain the coefficients and the equation can be used to predict the response.

The effects of the independent factors on the response variables in the separation process were assessed using pure error, considering a confidence level of 95% for all the variables. The effect of each factor and its statistical significance, for each of the response variables, were analyzed using ANOVA and standardized Pareto chart. The response surfaces of the respective mathematical models were also obtained, and their significance was accepted at $p \leq 0.05$. A multiple response optimization was carried out through the combination of experimental factors, to maximize the desirability function for the responses in the extracts. The desirability function method was applied to generate optimum conditions having some specific desirability value (close to =1 indicate that the setting achieves favorable results for all responses). All variables were obtained with the same equal weight = 1 (maximizing response).

2.4. Kinetic Study

A kinetic curve was plotted between the optimal extraction time versus the accumulated extract and total carotenoids content. The kinetic study was performed at the central point of the experimental design (30 MPa, 50 °C, and 4% co-solvent ethanol, v/v) as described by Gilbert-López et al. [19]. Each sample was analyzed for 4–6 min in the first 1 h and then 15 min for a total of 150 min. In this assay, the extraction yield (Y, %) and total carotenoids content (TCC, mg/g biomass) were calculated at each point of the curve. This assay was performed in duplicate with a total of 34 points per sample.

2.5. Extracts Analysis

2.5.1. Total Carotenoids Content (TCC)

Total carotenoids were determined in *I. galbana* biomass using a fucoxanthin standard (Sigma-Aldrich, 0–50 ppm) on a UV–Vis spectrophotometer (Shimadzu UV–1280, Kyoto, Japan). The absorption spectrum of total carotenoids was assessed in the maximum absorption wavelength selected of fucoxanthin because it is the main pigment on total carotenoids described in *I. galbana* [19,20]. This wavelength was 447.4 nm and Equation (2) was used to determine the total carotenoids content.

$$TCC = \frac{(A447.4 \times 8.66 \times DF \times V)}{M_{biomass}}$$

(2)

where TCC = total carotenoids content in mg/g biomass, A447.4 = the absorbance of the sample at λ max, 8.66 = the specific slope of the standard curve, DF = dilution factor of solvent, V = the solvent volume used in mL, and finally, Mbiomass is the mass of *I. galbana* in mg. Then, each ethanolic extraction after SFE of *I. galbana* biomass was measured and the final concentration of total carotenoids was calculated. The determination was carried out in triplicate (n = 3).

2.5.2. Total Carotenoids Recovery

The effect of operating conditions on total carotenoids extraction was expressed in terms of recovery, which was calculated on the basis of the initial mass of each compound as per Equation (3):

$$Recovery\ (\%) = \left(\frac{Wc}{Wt}\right) \times 100$$

(3)

where WC = mass of the compound extracted (mg); Wt = theoretical mass of the compound from a conventional extraction (mg). The total carotenoids were extracted by a conventional method using methanol for 24 h in a shaker incubator at 300 rpm at 30 °C, providing an average extracted of 24.4 ± 2.2 mg/g of total carotenoids.

2.5.3. Total Phenol Content (TPC)

Estimation of TPC was based on the 96-well microplate Folin–Ciocalteu method described by Ainsworth and Gillespie [21]. A total of 20 µL of the diluted extract (2.0 mg/mL) were mixed with 100 µL of 10% (v/v) Folin–Ciocalteu reagent and shaken. The mixture was left resting for 5 min. then 75 µL of sodium carbonate solution (700 mM) was added and was again shaken for 1 min. After 60 min at room temperature, the absorbance was measured at 765 nm on a microplate reader (BioTek Synergy HTX multi-mode reader, software Gen 5 2.0, Winooski, VT, USA). The absorbance of the same reaction with methanol, instead of the extract or standard, was subtracted from the absorbance of the reaction with the sample. For calibration, gallic acid dilutions (0–2 mg/mL) were used as standards. The results were expressed as gallic acid equivalents (GAE)/g biomass and were presented as the average of three measurements.

2.5.4. Determination of Antioxidant Activity

The Trolox equivalents antioxidant capacity (TEAC) value was determined using the method described by Re et al. [22], with a few modifications [23]. 2,2′-Azino-bis (3-ethylbenzothiazoline-6-sulfonic acid) diammonium salt (ABTS•+) radical was produced by reacting 7 mM ABTS and 2.45 mM potassium persulfate in the dark at room temperature for 16 h. The aqueous ABTS•+ solution was diluted with 5 mM sodium phosphate buffer at pH 7.4 to an absorbance of 0.7 (± 0.02) at 734 nm. Then, 20 µL of sample and 180 µL of ABTS•+ solution was added in a 96-well microplate reader of a spectrophotometer. The absorbance was measured at 734 nm within 10 min of the reaction. 6-hydroxy-2,5,7,8-tetramethylchromane-2-carboxylic acid (Trolox) was used as reference standard and results were expressed as TEAC values (mmol Trolox equivalents (TE)/g biomass). All analyses were done in triplicate (n = 3).

2.5.5. Extraction of Fatty Acid

The extraction of fatty acid methyl esters (FAMEs) was performed as per the direct acid catalysis method described by Lamers et al. [24] with a few modifications. Briefly, the reaction mixture containing

10 mg of SFE extract, 10 ppm of internal standard, and 3 mL of 5% (v/v) H_2SO_4 solution in methanol was incubated at 80 °C for 1 h with continuous agitation. Then, the flasks were washed with hexane and Milli-Q water until the pH of the water after washing was neutral. The mixture was separated into two layers by centrifugation (360 ×g, 10 min). The upper oil layer (FAMEs diluted in hexane) was separated and washed with Milli-Q water for further analysis and quantification by gas chromatography.

2.5.6. Analysis of Fatty Acids

In order to analyze the fatty acid composition, a gas chromatograph (Shimadzu 2010, Kyoto, Japan) equipped with a flame ionization detector (FID) and a split/splitless injector was used. In all the cases, samples (1 μL) were injected into a capillary column (RESTEK; 30 m, 0.32 mm i.d., 0.25 μm film thickness). The injector temperature was maintained at 250 °C in the split mode with a split ratio of 20:1 and nitrogen was used as the carrier gas at a constant flow rate of 1.25 mL/min. The oven temperature was maintained at 80 °C for 5 min, increased to 165 °C at the rate of 4 °C/min and maintained for 2 min, and further increased to 180 °C at the rate of 2 °C/min and maintained for 5 min. It was further heated at a rate of 2 °C/min to 200 °C for 2 min, then at a rate of 4 °C/min to 230 °C and maintained for 2 min, and finally maintained at that temperature for 2 min, reaching 250 °C at 2 °C/min. The detector temperature was 280 °C. Individual FAMEs were identified by comparing their retention times with those of mixed FAME standards (FAME Mix C4-C24, Supelco Analytical) and quantified by comparing their peak area with those of mixed FAME standards and an internal standard (tripentadecanoin ~10 ppm/sample, Nu-Check Pre, Inc., Elysian, MN, USA).

3. Results and Discussion

3.1. Specific Kinetics and Selection of Box-Behnken Design of Supercritical Fluid Extraction from Isochrysis galbana

Optimum extraction time was set by analyzing the kinetics of the extraction carried out by SFE. Figure 2 shows the evolution of the performance and the TC accumulated vs. extraction time of 150 min. The condition was marked as a central point of the experimental design chosen (30 MPa, 50 °C, 4% v/v of co-solvent ethanol). Each extract was recovered at the established time point as described in the material and methods section and the percentage of the extractable was calculated. The yield extraction curve reached a plateau at 60 min and total carotenoids between 100 and 120 min.

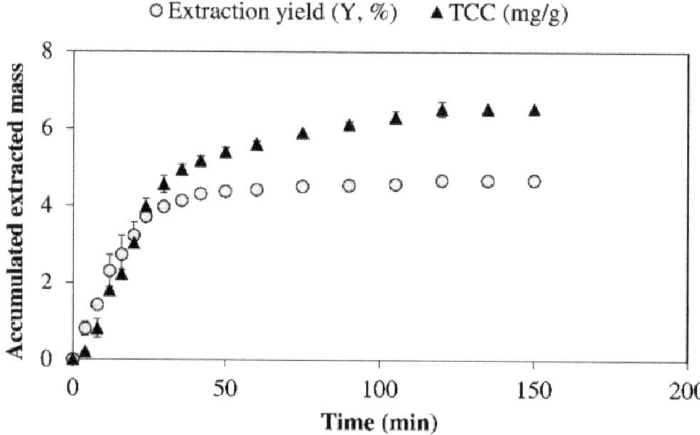

Figure 2. Kinetics of improving extraction yield (Y, %) and total carotenoids content (TCC) from *I. galbana* during 150 min. The working conditions were at the central point: 50 °C, 30 MPa, 4% co-solvent (ethanol, v/v), and a CO_2 flow rate of 7.2 g/min.

For consolidating the maximum in all the variables studied, an optimal extraction time of 120 min was set for each SFE condition as after this time there was no increase in the amount of extracted material. In particular, the maximum value of total carotenoids recollected was approximately 99.80% (~ 6.53 mg/g) and the extraction yield was 99.8% (~ 4.67%, w/w) at 120 min.

On the other hand, Box-Behnken statistical design was selected for this study because previous works demonstrated that models 2 K did not represent adequately the bioactive compounds extraction process (results not shown). Respect to the percentage of co-solvent, we also tested with less range (0%-4% v/v) and all variables content was improved under 4% ethanol. For this reason, we expand to 8% of the volume of ethanol in order to minimize the cost process under the Box-Behnken design. Table 1 shows the results of the ANOVA and Regression coefficients for each variable. Here, R-Square statistic indicated that the quadratic model was well adjusted in against lineal or second order. It means the variability in all studied variables. In the case of the adjusted R-square statistic were slightly smaller than R-squared and in the same way lack-of-fit determined that the selected model was adequate to describe the observed data. This test was performed by comparing the variability of the current model residuals to the variability between observations at replicate values of the independent variable. Since the *p*-value for lack-of-fit in the ANOVA table was greater or equal to 0.05, the model appears to be adequate for the observed data (except FAME data that was close to 0.05).

Table 1. Regression coefficients (values of variables are specified in their original units, extraction yields (Y), total carotenoids content (TCC), TC recovery, total phenol content (TPC), Trolox equivalents antioxidant capacity (TEAC) antioxidant method), fatty acid methyl ester (FAMEs), and statistics for the fit obtained by multiple linear regression.

Terms of the Model	Y		TCC		TC recovery		TPC		TEAC		FAMEs	
	Estimated	*p*-value	Estimated	*p*-value	Estimated	*p*-value	Estimated	*p*-value	Estimated	*p*-value	Estimated	*p*-value
constant	51.47		35.73		146.47		−619.40		−2.85		−17.72	
A:P	−0.09	0.05 *	0.11	0.18	0.44	0.18	2.38	0.32	−0.0007	0.02*	0.07	0.15
B:T	−1.44	0.14	−1.73	0.67	−7.10	0.67	17.03	0.19	0.124	0.81	0.44	0.43
C:Co-solvent	−1.23	0.0012 *	−2.27	0.009*	−9.32	0.009 *	−21.94	0.07	0.0364	0.05 *	−0.63	0.01 *
AA	0.00002	0.83	−0.00021	0.31	−0.00086	0.31	−0.0045	0.05 *	$-3.33\,E^{-7}$	0.91	−0.0001	0.06
AB	0.0019	0.051	0.0004	0.83	0.0016	0.84	0.0030	0.87	0.00003	0.29	0.0004	0.64
AC	0.0001	0.96	0.0049	0.33	0.020	0.33	0.0775	0.12	−0.00001	0.86	0.0002	0.91
BB	0.0078	0.37	0.014	0.48	0.058	0.48	−0.2061	0.29	−0.00136	0.01 *	−0.0064	0.17
BC	0.045	0.06	0.037	0.45	0.151	0.45	0.2357	0.59	0.00069	0.36	0.032	0.16
CC	−0.025	0.63	0.033	0.79	0.136	0.79	−0.7877	0.50	−0.00677	0.01 *	−0.061	0.10
Lack-of-Fit		0.13		0.29		0.30		0.10		0.05		0.04
Statistics for the goodness of fit of the model												
R^2	0.930		0.825		0.825		0.806		0.918		0.874	
Adjusted R^2	0.804		0.510		0.510		0.457		0.770		0.647	
RSD	1.506		3.565		14.616		33.068		0.054		1,567	
P	0.336		0.905		0.905		0.356		0.871		0.251	
C.V.	0.639		0.647		0.647		0.676		0.529		0.558	

Note: R^2—determination coefficient, adjusted R^2, RSD—residual standard deviation, *p*-value of the lack-of-fit test for the model; C.V.—coefficient of variation; * significant coefficients of the model.

3.2. Effects of Different Parameters on the Extraction Yield and Total Carotenoids Content and Recovery

The experimental conditions and results of the Box-Behnken design for the extraction conditions for *I. galbana* by SFE for 120 min are listed in Table 2. The range of extraction yield (Y) was 1.09–12.82% (w/w) and that of total carotenoids content (TCC) was 1.34–19.01 mg/g.

In this table, the total carotenoids recovery is also mentioned, which was estimated as the percentage of the total carotenoids extracted in each sample to the total carotenoids extracted in a conventional methanol extraction, which was assumed to be 100% (24.40 ±2.24 mg/g) with respect to dry biomass. The recovery of total carotenoids reached 5.50% under pure CO_2 and 77.93% when CO_2 modified with 8% (v/v) ethanol was used. The conditions were at 20 MPa/50 °C and at 300 MPa/60 °C, respectively.

Letters in the experiment column are the acronyms of the tested variables: Pressure (P) and Temperature (T). Values are represented as a mean standard deviation. It was SD ≤ 5% (n = 3 analytical measurement). All values were calculated per gram of initial biomass.

Table 2. Extraction yields (Y), total carotenoids content (TCC), total carotenoids recovery (TC recovery), total phenol content (TPC), TEAC antioxidant method, and fatty acid methyl ester (FAMEs) by SFE from freeze-dried *Isochrysis galbana* using Box-Behnken experimental design. The general parameters were biomass loading = 2.0 g, CO_2 flow rate = 7.4 g/min, extraction time = 120 min.

Run	P (MPa)	T (°C)	Co-solvent (%)	Y (%, w/w)	TCC (mg/g)	TC Recovery (%, w/w)	TPC (mg GAE/g)	TEAC (mmol TE/g)	FAMEs (mg/g)
1	30	40	8	5.71 ± 0.24	14.09 ± 0.55	57.75 ± 2.30	93.33 ± 3.52	$0.11 \pm 5.3 \times 10^{-3}$	3.41 ± 0.13
2	40	40	4	6.16 ± 0.16	9.66 ± 0.32	39.61 ± 1.45	50.93 ± 2.01	$0.28 \pm 1.3 \times 10^{-2}$	5.56 ± 0.16
3	40	60	4	10.25 ± 0.49	6.22 ± 0.29	25.49 ± 1.10	22.89 ± 1.02	$0.33 \pm 1.4 \times 10^{-2}$	5.41 ± 0.21
4	40	50	0	2.28 ± 0.11	4.05 ± 0.16	16.61 ± 0.74	5.98 ± 0.23	$0.22 \pm 1.1 \times 10^{-2}$	1.18 ± 0.04
5	20	50	0	1.09 ± 0.03	1.34 ± 0.06	5.50 ± 0.21	5.71 ± 0.20	$0.15 \pm 7.5 \times 10^{-3}$	0.47 ± 0.02
6	30	50	4	5.79 ± 0.21	7.02 ± 0.33	28.78 ± 1.35	109.31 ± 3.56	$0.31 \pm 1.4 \times 10^{-2}$	7.57 ± 0.26
7	40	50	8	8.78 ± 0.32	15.33 ± 0.72	62.85 ± 2.68	157.16 ± 3.66	$0.31 \pm 1.5 \times 10^{-2}$	7.19 ± 0.33
8	30	50	4	5.19 ± 0.24	10.83 ± 0.51	44.38 ± 1.89	94.40 ± 4.02	$0.40 \pm 1.8 \times 10^{-2}$	7.63 ± 0.29
9	30	50	4	4.36 ± 0.20	6.00 ± 0.28	24.57 ± 1.02	120.77 ± 4.02	$0.33 \pm 1.5 \times 10^{-2}$	6.87 ± 0.32
10	30	60	8	12.82 ± 0.06	19.01 ± 0.92	77.93 ± 2.87	76.06 ± 3.52	$0.20 \pm 9.0 \times 10^{-3}$	8.82 ± 0.38
11	30	60	0	1.64 ± 0.08	2.74 ± 0.18	11.24 ± 0.42	37.71 ± 1.44	$0.04 \pm 1.0 \times 10^{-3}$	2.86 ± 0.12
12	20	40	4	5.73 ± 0.12	9.11 ± 0.45	37.33 ± 1.63	67.87 ± 3.21	$0.15 \pm 6.2 \times 10^{-3}$	3.50 ± 0.13
13	20	50	8	7.43 ± 0.35	4.85 ± 0.23	19.87 ± 0.75	32.90 ± 1.20	$0.26 \pm 1.2 \times 10^{-2}$	6.09 ± 0.28
14	30	40	0	1.80 ± 0.08	3.72 ± 0.17	15.25 ± 0.65	92.69 ± 3.54	$0.06 \pm 2.1 \times 10^{-3}$	2.63 ± 0.11
15	20	60	4	2.15 ± 0.10	4.10 ± 0.20	16.80 ± 0.82	28.04 ± 1.32	$0.07 \pm 2.9 \times 10^{-3}$	1.78 ± 0.07

The obtained mathematical models that maximize the yield (4), TCC (5), and TC recovery (6) were studied. The significant variables in the study, and which responded to the combined relationships between process variables, are given as follows:

$$Y = 51.46 - 0.093 \cdot P - 1.23 \cdot \text{Co-solvent} + 0.000018 \cdot P^2 + 0.0019 \cdot P \cdot T + 0.0001 \cdot P \cdot \text{Co-solvent} + 0.045 \cdot T \cdot \text{Co-solvent} - 0.025 \cdot \text{Co-solvent}^2 \tag{4}$$

$$TCC = 35.73 - 2.27 \cdot \text{Co-solvent} + 0.0049 \cdot P \cdot \text{Co-solvent} + 0.037 \cdot T \cdot \text{Co-solvent} + 0.033 \cdot \text{Co-solvent}^2 \tag{5}$$

$$TC\ recovery = 146.47 - 9.32 \cdot \text{Co-solvent} + 0.020 \cdot P \cdot \text{Co-solvent} + 0.15 \cdot T \cdot \text{Co-solvent} + 0.14 \cdot \text{Co-solvent}^2 \tag{6}$$

Table 1 further provides more information on the obtained statistical results by presenting the estimated regression coefficients of all the factors and interactions for each response variable.

In the case of the variables TCC content and TC recovery (Equations (4) and (5)), only the co-solvent factor was found significant in the extraction process. For the response variable Y (Equation (3)), the pressure factor was added to the co-solvent factor, as both were relevant from of *I. galbana* using SFE. In addition, Figure 3A to 3C show the response surfaces based on the selection of factors for the best recoveries of Y, TCC, and TC recovery, respectively. As is evident, higher Y, TCC, and TC recovery were obtained with an increase in co-solvent percentage.

This corroborates with the outcomes described by Gilbert-López et al. [19]. This study demonstrated that sequential steps using various pressurized green solvents have improved selectivity for bioactive compound recovery from *Isochrysis galbana*. The assays performed by SFE under pure CO_2 improved the yield of triacylglycerides but not carotenoids (optimal condition was 30 MPa/50 °C, achieving 16.2 ± 0.3 mg carotenoids/g extract and 5% (w/w) of yield). Conversely, the best possible recovery of fucoxanthin was observed by increasing the ethanol co-solvent to 45% (v/v), denominated as carbon dioxide expanded ethanol (CXE) extraction under low pressures [19]. This recovery of carotenoids is in accordance with a study by Conde et al. [25] using ethanol as a modifier and allowed higher extraction yields, particularly of fucoxanthin.

The yield also improved when the pressure was increased to 40 MPa. Several reports confirm that high pressure and temperature result in better yield for the same extraction duration [25,26]. A clear benefit of the increased pressure with co-solvent was the enhanced solvent density of CO_2. Nevertheless, the temperature was not a significant parameter in the results of our statistical analyses, but in general, the best response was observed under higher values of this factor such as 60 °C

(Figure 3). In this way, the "optimal values" delivered by the software for statistical analysis were 15.15%, 18.14 mg/g, and 74.36% w/w of Y, TCC, and TC recovery, respectively, under 40 MPa, ~60 °C and ~8% co-solvent. These experiences, in a certain way corresponded to the extraction conditions used in the experiment where the variables were improved (30 MPa, 60 °C, and 8.0% co-solvent) obtaining a lower value of 12.82% in extraction yield. However, TCC values and TC recovery of 19.01 mg/g and 77.93% w/w, respectively, were slightly higher than those predicted by the software.

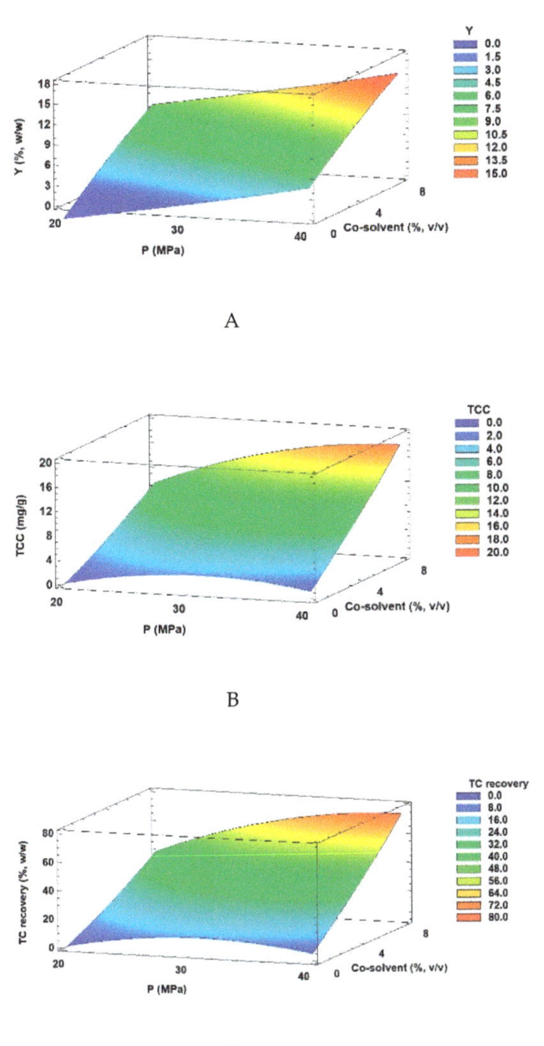

A

B

C

Figure 3. Response surface plot showing combined effects of pressure and co-solvent on (**A**) extraction yield (Y), (**B**) total carotenoids content (TCC), and (**C**) recovery (TC recovery) from *I. galbana*. All extractions were done at 60 °C temperature.

In general, the values achieved in this study are better than other reports of specific carotenoids extraction from natural sources. For example, the microalgae *Phaeodactylum tricornutum* or *Odontella aurita* [27,28] gave higher yield than those by brown seaweeds including *Laminaria japonica*, *Eisenia bicyclis*, and *Undaria pinnatifida* [27,29,30]. Kim et al. [27] reported several methods and solvents to maximize fucoxanthin extraction from *P. tricornutum*, achieving the best yield of 16.51 mg/g under pressurized liquid extraction (PLE) at 10.3 MPa, 100 °C, 100% ethanol, and static time of 30 min. The content yielded was similar to that by maceration extraction (15.71 mg/g). Likewise, *O. aurita* is another microalga and a natural producer of fucoxanthin as described by Xia et al. [28]. In their study, besides optimizing culture conditions for improving this carotenoid, they attempted to determine the optimal conditions for fucoxanthin extraction using five conventional solvents. The best yield of fucoxanthin among the five tested solvents was obtained with methanol (16.18 mg/mg), followed by ethanol (15.83 mg/g) and acetone (13.93 mg/g). Generally, conventional methods are applied to extract bioactive compounds (especially from carotenoids family) using organic solvents [31] achieving better extraction yield although they are often required large volumes, making the method expensive and environmentally unfriendly [32]. For instance, Kim et al. [20] used different solvents and extraction time to improve the fucoxanthin recovery and were able to obtain 20.87 mg/g of fucoxanthin from dry biomass of marine microalga *I. aff. galbana* (CCMP1324) using acetone. These results are better than total carotenoids rich in fucoxanthin extracted by SFE in our study, which may be due to an increase in solvent density and swelling of the matrix because of co-solvent addition.

3.3. Total Phenolic Content and Antioxidant Response in Isochrysis galbana by Supercritical Fluid Extraction

Table 2 presents the results of total phenolic content (TPC) and TEAC method as an antioxidant response. The estimated response surfaces are described in Figure 4A based on the parameters for TPC, which was complemented with a summarized second-order polynomial Equation (7) as a mathematical model to find the optimum conditions that maximize TPC as follows:

$$\text{TPC} = -619.40 + 2.38 \cdot \text{P} - 0.0045 \cdot \text{P}^2 + 0.0029 \text{P} \cdot \text{T} + 0.077 \cdot \text{P} \cdot \text{Co-solvent} \tag{7}$$

This quadratic interaction of the pressure factor was significant for this response variable (TPC) as is evident in Table 1. The optimal condition was determined to be 34.8 MPa, 48 °C, and 8% co-solvent with an optimal yield of 133.9 mg GAE/g biomass. These results are better than those reported in other studies using *I. galbana* as natural biomass rich in bioactive compounds through conventional methods of extraction [33,34]. For example, Widowati et al. [33] and Foo et al. [35] reported yields from *I. galbana* clone Tahiti, and *I. galbana* to be nearly 17.79 mg GAE/g and 12.24 mg GAE/g, respectively. Many reports compare SFE with conventional processes because they have advantages and disadvantages in both cases. In particular, fluids under supercritical condition reducing process time and enhancing the extraction yield. Parameters as temperature or pressure ease selectivity, effective penetrating the biomass and better mass transfer between phases versus conventional methods [18,36].

The results mentioned in Table 2 indicate enhanced TPC content with an increase in the percentage of co-solvent (8% under optimal condition). In algae, the use of co-solvent such as ethanol increases the polarity of SFE and with it, the phenolic compounds, antioxidant activity, and fucoxanthin recovery are expected to increase [25,37,38]. *Sargassum muticum* is a brown alga rich in fucoxanthin, fatty acids and phenolic compounds described by Conde et al. [25]. They demonstrated that an increase in the ethanol concentration strongly improved variables such as total yield, radical scavenging capacity, and the fucoxanthin extraction yield. However, due to the lower process selectivity, the phenolic content using maximum co-solvent (10% ethanol) was moderately higher than with pure CO_2. Other reports of TPC extraction from conventional methods include studies by Goiris et al. [39] on *Isochrysis* ISO-T and *Isochrysis* sp. giving yields of 2.67 and 4.57 mg GAE/g, respectively, which are relatively low. In a study performed by Li et al. [40], 23 microalgae were evaluated using sequential organic solvent extraction and it was observed that *Nostoc ellipsosporum* CCAP 1453/17 had a TPC of 60.35 ± 2.27 mg GAE/g. The antioxidants, such as fucoxanthin and phenolic compound in *Isochrysis* sp., have been prescribed for the prevention of the age-related changes in the central nervous system as they scavenge free radicals and reactive oxygen species (ROS) [41].

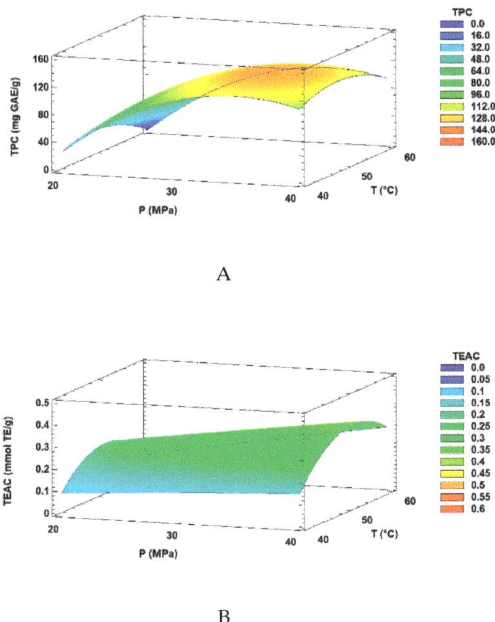

A

B

Figure 4. Response surface plot showing combined effects of pressure and temperature on (**A**) phenol content (TPC) and (**B**) TEAC method from *I. galbana*. All extractions were done with 8% co-solvent (ethanol, v/v).

The use of ABTS in the radical scavenging assay is a popular indirect method for determining the antioxidant capacity of bioactive compounds [42,43]. The antioxidant response was determined by TEAC method. The mathematic model can be described as the second-order polynomial Equation (8):

$$TEAC = -2.85 - 0.00074{\cdot}P + 0.036{\cdot}\text{Co-solvent} - 3.33E^{-7}{\cdot}P^2 + 0.000032{\cdot}P{\cdot}T - 0.000012{\cdot}P{\cdot}\text{Co-solvent} - 0.0014{\cdot}T^2 + 0.00069{\cdot}T{\cdot}\text{Co-solvent} - 0.0068{\cdot}\text{Co-solvent}^2 \tag{8}$$

Here, several factors important in the TEAC determination such as pressure, presence of co-solvent, and quadratic interaction of temperature and co-solvent in SFE extracts of *I. galbana*. Figure 4B shows the estimated response surface and the obtained TEAC values are described in Table 2. The values were between 0.04 and 0.40 mmol TE/g at SFE condition of 30 MPa, 60 °C, and 0% co-solvent and 40 MPa, 60 °C, and 4% co-solvent, respectively. The optimal value was 0.42 mmol TE/g at 40 MPa, 51.6 °C, and 4.9% co-solvent. Likewise, a study by Reyes et al. [44] demonstrated a similar influence of the factors on the TEAC response from *H. pluvialis*.

Particularly, our results indicate the influence of the solvent on the yield. Other authors have also described the possibility of using different combinations of ethanol + water as co-solvent for green extraction of bioactive compounds from a diverse natural source [45–47]. The inclusion of water as co-solvent has been described as advantageous because it influences a fast and quantitative recovery of the phenolic compounds, high anthocyanin concentration, and high antioxidant capacity [46,48]. Moreover, a reduced percentage of ethanol minimizes the cost and impact on the environment [48].

The antioxidant activity in this study was higher than that reported by other conventional methods. For example, Goiris et al. [39] described *Tetraselmis* sp. with the best antioxidant capacity among 23 microalgae at 69.4 µmol TE/g. *I. galbana* is rich in antioxidant activity because of the presence of carotenoids. The number of double and allenic bonds and the presence of an acetyl functional groups in fucoxanthin are also responsible for the higher antioxidant activities [42,49]. Fucoxanthin and its derived metabolites display antioxidant activities comparable to that of α-tocopherol [43]. Nevertheless, very limited information is available about the phenolic contents and antioxidant activity using SFE in *I. galbana*.

3.4. Measurement of Fatty Acid Composition

The fatty acid profiles were also identified along with FAME content (mg FAME/g) because they can be used in different biotechnological applications. The FAME content per gram of biomass is listed in Table 2. The range of FAME content was from 0.47 to 8.82 mg/g and Run 10 (30 MPa, 60 °C, and 8% co-solvent) gave the best results of 8.82 mg/g followed by the central point of the Box-Behnken design with values in the range of 6.87–7.63 mg/g. A summarized polynomial equation was obtained to establish a mathematical model, reach optimum operating conditions, and maximize FAMEs contents from *I. galbana* by SFE. Equation (9) and data in Table 1 indicate the relevance of the co-solvent factor ($p < 0.05$) in the FAME quantification under these SFE conditions.

$$\text{FAME} = -39.50\ 5 - 0.30 \cdot \text{Cosolvent} + 0.0002 \cdot P \cdot \text{Cosolvent} + 0.03 \cdot T \cdot \text{Cosolvent} - 0.10 \cdot \text{Cosolvent}^2 \tag{9}$$

The optimal value suggested for the response work surface was 8.84 mg/g under the experimental condition of 33.3 MPa, 57 °C, and 7.9% co-solvent, nearly similar to that observed in our study. These results are presented in Figure 5 (response surface graphic) where at medium pressure, with high temperature and co-solvent, FAME content increased in *I. galbana* although the co-solvent was the only significant factor in the process. Extractions under increased pressures gave the best results as the solubility of triglycerides could be improved due to an increase in the solvent density.

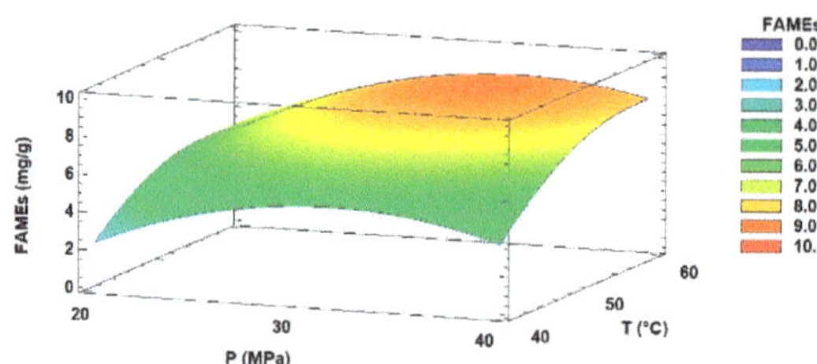

Figure 5. Response surface plot showing combined effects of pressure and temperature on FAMEs content from *I. galbana*. All extractions were done with 8% co-solvent (ethanol, v/v).

Table 3 presents the profile of primary fatty acids derived from SFE extracts of *I. galbana*. The profile (measured as % area) consists of polyunsaturated fatty acids (PUFAs) in the range of 25.4–95.6%, MUFAs (monounsaturated fatty acids) in the range of 0.0–47.7%, and saturated fatty acids (SFAs) in the range of 4.4–42.1% of total fatty acids. In general, the fatty acid profile was highlighted by linoleic acid (C18:2) followed by linolenic acid (C18:3). Myristic acid (C14:0) and palmitic acid (C16:0) were the main SFAs. The highest recoveries of MUFAs and PUFAs were found under two conditions at 20 MPa, 40 °C, and 4% co-solvent (Run 12) and 30 MPa, 50 °C, and 4% ethanol (Run 9), at nearly 47.7 and 95.6%, respectively. SFE is considered an appropriate method for the extraction of fatty acids and lipids or for compounds of low polarity from microalgae because of its non-polar property [50]. It is also a solvent selective for neutral lipids such as triglycerides but does not solubilize phospholipids [51]. Our results corroborate with several reports about the extraction of polar compounds, which could be enhanced by adding of polar co-solvents, such as ethanol [52,53].

Table 3. Fatty acid composition and content of SFE extracts from freeze-dried *Isochrysis galbana* used in this study (% area of total FAME and mg FAME/g). The general parameters were biomass loading = 2.0 g, CO_2 flow rate = 7.4 g/min, extraction time = 120 min (SD ≤ 5%; n = 3).

Fatty Acid	Common Name	Run														
		1	2	3	4	5	6	7	8	9	10	11	12	13	14	15
C14:0	Myristic acid	20.0 ± 0.5	n.d.	n.d.	22.8 ± 1.02	4.9 ± 0.21	n.d.	n.d.	3.7 ±0.2	n.d.	24.2 ±1.2	n.d.	0.2 ± 0.01	n.d.	n.d.	n.d.
C16:0	Palmitic acid	n.d.	4.9 ± 0.2	13.8 ± 0.5	9.1 ± 0.40	6.6 ± 0.28	5.6 ± 0.1	n.d.	7.1 ±0.3	n.d.	n.d.	n.d.	7.8 ± 0.3	8.5 ±0.38	11.0 ±0.4	9.7 ± 0.4
C16:1	Palmitoleic acid	9.8 ± 0.3	n.d.	n.d.	n.d.	n.d.	9.0 ± 0.3	8.9 ±0.3	14.9 ±0.5	n.d.	30.1 ±1.3	3.0 ±0.1	n.d.	n.d.	7.4 ±0.3	n.d.
C18:0	Stearic acid	n.d.	n.d.	n.d.	n.d.	n.d.	n.d.	n.d.	2.4 ±0.06	n.d.	2.4 ±0.05	3.1 ±0.1	1.7 ± 0.08	1.4 ±0.06	1.3 ±0.04	1.4 ± 0.07
C18:1	Oleic acid	n.d.	n.d.	n.d.	n.d.	2.2 ± 0.09	0.6 ± 0.02	n.d.	2.3 ±0.10	n.d.	n.d.	n.d.	47.7 ±1.7	36.7 ±1.5	37.9 ±1.7	36.2 ± 1.5
C18:2	Linoleic acid	39.1 ± 1.7	43.5 ± 1.9	6.0 ± 0.3	2.7 ± 0.11	35.0 ± 1.42	48.1 ± 1.7	41.9 ± 1.8	40.7 ±1.8	53.6±1.68	43.3 ±1.8	88.1 ±3.5	26.1 ±1.2	17.9 ±0.7	18.9 ±0.7	18.2 ± 0.8
C18:3	Linolenic acid	26.2 ± 1.2	36.5 ± 1.7	54.1 ± 2.5	48.4 ± 1.89	28.7 ± 1.25	25.5 ± 1.2	24.9± 1.0	n.d.	42.0 ±1.89	n.d.	0.9 ±0.04	4.0 ±0.1	19.6 ±0.8	2.9 ±0.1	n.d.
C20:0	Eicosanoic acid	n.d.	n.d.	5.8 ± 0.2	3.5 ± 0.17	3.9 ± 0.15	1.6 ± 0.04	2.9±0.05	1.5 ±0.06	n.d.	n.d.	4.1 ±0.2	10.2 ±0.5	12.6 ±0.5	15.3 ±0.7	22.6 ± 1.1
C21:0	Methyl heneicosanoate	n.d.	n.d.	n.d.	n.d.	10.8 ± 0.42	n.d.	9.2±0.3	8.8 ±0.3	n.d.	n.d.	n.d.	n.d.	n.d.	n.d.	n.d.
C22:2	Cis-13,16-docosadienoic acid methyl ester	n.d.	n.d.	n.d.	n.d.	n.d.	n.d.	n.d.	n.d.	n.d.	n.d.	n.d.	2.2 ±0.1	3.4 ±0.1	3.6 ± 0.1	8.2 ± 0.3
C20:3	Cis-11-14-17-eicosatrienoic acid methyl ester	5.1 ± 0.2	8.5 ± 0.3	11.5 ± 0.5	7.8 ± 0.3	6.0 ± 0.3	6.2 ± 0.2	n.d.	n.d.	n.d.	n.d.	n.d.	n.d.	n.d.	n.d.	n.d.
Others		n.d.	6.5 ± 0.3	8.8 ± 0.4	5.8 ± 0.2	1.9 ± 0.08	3.4 ± 0.1	12.2± 0.6	18.6 ±0.8	4.4 ±0.1	n.d.	n.d.	n.d.	n.d.	1.8 ±0.0	3.8 ± 0.1
Σ SFAs	Total saturated fatty acids	20.0	6.7	23.3	35.4	26.2	7.2	24.3	42.1	4.4	26.6	7.1	20.0	22.5	29.3	37.5
Σ MUFAs	Total monounsaturated fatty acids	9.8	4.8	5.2	5.8	4.1	13.0	8.9	17.2	n.d.	30.1	3.0	47.7	36.7	45.3	36.2
Σ PUFAs	Total polyunsaturated fatty acids	70.3	88.5	71.6	58.8	69.7	79.8	66.8	40.7	95.6	43.3	89.9	32.4	40.9	25.4	26.4

Note: SFE = supercritical fluids extraction; FAME = fatty acid methyl ester; n.d. = no detection. (SD ≤ 5% ; n = 3).

Machmudah et al. [54] reported that an increase in the extraction pressure causes intensification in carbon dioxide density and, consequently, an increase in the solvation power for fatty acids. Our results show that under low pressures, more saturated fatty acids are extracted and when the pressure is increased, the proportion of unsaturated fatty acids increases in the extracted phase, an observation also supported by Cheung [55]. In general, this indicated that the triglycerides containing the more unsaturated fatty acids could be soluble at higher densities of CO_2 due to increase pressure factor. However, the combined effects of pressure and temperature on the overall solubility of PUFA can vary because it depends on their chain length and there seems to be a compromise between supercritical fluid density and vapor pressure of the solute concerned [55].

3.5. Desirability Function

The desirability to maximize all studied variables has been summarized in Figure 6. This procedure helped in determining the combination of experimental factors that involve simultaneous optimization of several response variables and was selected for meeting these goals and giving equal importance to all responses. The maximum 'desirability' predicted by the software was obtained in Run 7 with parameters of 40 MPa, 50° C, and 8% co-solvent. The results were obtained as the extraction yield of 8.78%, TCC of 15.33 mg/g, TC recovery of 62.85%, TPC of 157.16 mg GAE/g, TEAC of 0.31 mmol TE/g, and FAME of 7.19 mg/g. The optimum desirability was obtained at 38.4 MPa, 56.74 °C, and 8% co-solvent where the optimal results were extraction yield of 12.82%, TCC of 16.96 mg/g, TC recovery of 69.53%, TPC of 114.57 mg GAE/g, TEAC of 0.32 mmol TE/g, and FAME of 8.31 mg/g. Finally, the optimization desirability was determined to be 0.86 and the experiments under the optimum conditions provided values close to those predicted by the statistical model.

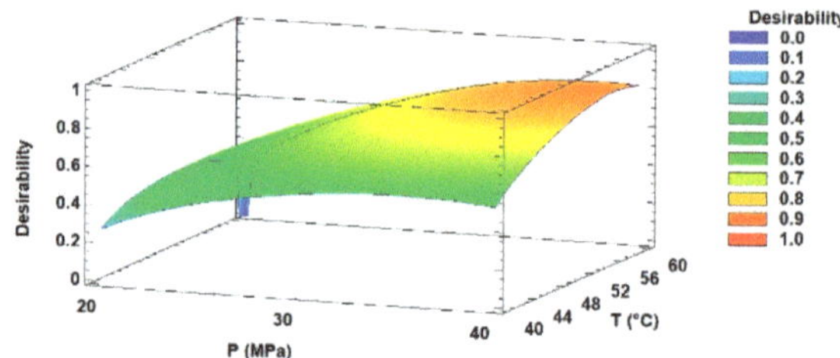

Figure 6. Surface of the desirability function in terms of pressure, temperature and maximum co-solvent (8% ethanol, v/v) obtained for maximizing extraction yield (Y), total carotenoids content (TCC) and recovery (TC recovery), phenol content (TPC), TEAC method, and FAMEs content from *I. galbana*.

4. Conclusions

The present study focused on the prospecting of bioactive compounds present in the microalgae *I. galbana* by SFE for potential use in food, pharmaceuticals, and cosmetics. Several variables were studied including extraction yield, total carotenoids content and recovery, total phenols, antioxidant activity (TEAC method), and fatty acid content and profile. In general, the addition of ethanol as co-solvent significantly increased the efficiency of SFE extraction for all variables, and to a lesser extent, TEAC as antioxidant measurement. Another parameter to enhance the bioactive compound extraction from *I. galbana* was pressure, although it was not a significant variable except in the TEAC method. Particularly, linoleic acid (C18:2) followed by linolenic acid (C18:3) was highlighted in the profile that was improved under higher pressure, thus indicating that as pressure increased, solubility of triglycerides containing the more unsaturated fatty acids increased at higher densities. Concurrently, desirability function that aided in the operational condition closer to the optimum was found in Run 7 (40 MPa, 50 °C, and 8% co-solvent, desirability predicted 0.79). To better explore the potential of *I. galbana*, new and improved supercritical extraction with a higher percentage of ethanol could be developed in spite of it could increase the operational costs in the process. In conclusion, this study provides functional information to optimize extraction of bioactive compounds from *I. galbana* using SFE because of its rich content of bioactive compounds and antioxidant activity with increasing demand in food, pharmaceuticals, and cosmetics market.

Author Contributions: Conceptualization, M.C.R.-D.; data curation, F.S. and E.M.; formal analysis, M.C.R.-D. and E.M.; funding acquisition, M.C.R.-D. and P.C.; investigation, M.C.R.-D., F.S., and E.M.; methodology, M.C.R.-D. and P.C.; project administration, M.C.R.-D. and P.C.; resources, G.R.-C.; software, M.C.R.-D. and P.C.; writing—original draft, M.C.R.-D.; writing—review and editing, P.C.. All authors have read and agreed to the published version of the manuscript.

Funding: This research was financed by several projects with public funds of Chile, grants FONDECYT-11170017, PAI-79160037 and FONDEQUIP EQM-160073 (CONICYT) and MINEDUC-UA project, code ANT-1855.

Acknowledgments: The research group "LAMICBA" at University of Antofagasta, Chile, thanks "Microalgas Oleas de México S.A." for providing microalgal samples.

Conflicts of Interest: There is no conflict of interests.

References

1. Nicoletti, M. Microalgae nutraceuticals. *Foods* **2016**, *5*, 54. [CrossRef] [PubMed]

2. Pawar, K.; Thompkinson, D.K. Multiple Functional Ingredient Approach in Formulating Dietary Supplement for Management of Diabetes: A Review. *Crit. Rev. Food Sci. Nutr.* **2014**, *54*, 957–973. [CrossRef] [PubMed]

3. Gantar, M.; Svirčev, Z. Microalgae and Cyanobacteria: Food for thought (1). *J. Phycol.* **2008**, *44*, 260–268. [CrossRef] [PubMed]

4. Chacón-Lee, T.; González-Mariño, G.E. Microalgae for "healthy" foods—Possibilities and challenges. *Compr. Rev. Food. Sci. Food Saf.* **2010**, *9*, 655–675. [CrossRef]

5. Gouveia, L.; Coutinho, C.; Mendonça, E.; Batista, A.P.; Sousa, I.; Bandarra, N.M.; Raymundo, A. Functional biscuits with PUFA-ω3 from *Isochrysis galbana*. *J. Sci. Food Agric.* **2008**, *88*, 891–896. [CrossRef]

6. Dufossé, L.; Galaup, P.; Yaron, A.; Arad, S.M.; Blanc, P.; Murthy, K.N.C.; Ravishankar, G.A. Microorganisms and microalgae as sources of pigments for food use: A scientific oddity or an industrial reality? *Trends Food Sci. Technol.* **2005**, *16*, 389–406. [CrossRef]

7. Koyande, A.K.; Chew, K.W.; Rambabu, K.; Tao, Y.; Chu, D.T.; Show, P.L. Microalgae: A potential alternative to health supplementation for humans. *Food Sci. Hum. Wellness* **2019**, *8*, 16–24. [CrossRef]

8. Pernet, F.; Tremblay, R.; Demers, E.; Roussy, M. Variation of lipid class and fatty acid composition of Chaetoceros muelleri and Isochrysis sp. grown in a semicontinuous system. *Aquaculture* **2003**, *221*, 393–406. [CrossRef]

9. Da Silva, R.P.; Rocha-Santos, T.A.; Duarte, A.C. Supercritical fluid extraction of bioactive compounds. *Trac-Trends Anal. Chem.* **2016**, *76*, 40–51. [CrossRef]

10. Guedes, A.C.; Amaro, H.M.; Malcata, F.X. Microalgae as sources of high added-value compounds—A brief review of recent work. *Biotechnol. Prog.* **2011**, *27*, 597–613. [CrossRef]

11. Mulders, K.J.; Weesepoel, Y.; Lamers, P.P.; Vincken, J.-P.; Martens, D.E.; Wijffels, R.H. Growth and pigment accumulation in nutrient-depleted *Isochrysis aff. galbana* T-ISO. *J. Appl. Phycol.* **2013**, *25*, 1421–1430. [CrossRef]

12. Kalam, S.; Gul, M.Z.; Singh, R.; Ankati, S. Free radicals: Implications in etiology of chronic diseases and their amelioration through nutraceuticals. *Pharmacologia* **2015**, *6*, 11–20.

13. Miyashita, K. Function of marine carotenoids. In *Food Factors for Health Promotion*; Karger Publishers: Kyoto, Japan, 2009; Volume 61, pp. 136–146.

14. Raposo, M.; de Morais, A.; de Morais, R. Carotenoids from marine microalgae: A valuable natural source for the prevention of chronic diseases. *Mar. Drugs* **2015**, *13*, 5128–5155. [CrossRef]

15. Sousa, I.; Gouveia, L.; Batista, A.P.; Raymundo, A.; Bandarra, N.M. *Microalgae in Novel Food Products*; Food Chemistry Research Developments: New York, NY, USA, 2008; pp. 75–112.

16. Simopoulos, A.P. The importance of the ratio of omega-6/omega-3 essential fatty acids. *Biomed. Pharmacother.* **2002**, *56*, 365–379. [CrossRef]

17. Gałuszka, A.; Migaszewski, Z.; Namieśnik, J. The 12 principles of green analytical chemistry and the significance mnemonic of green analytical practices. *Trac-Trends Anal. Chem.* **2013**, *50*, 78–84. [CrossRef]

18. Herrero, M.; Cifuentes, A.; Ibañez, E. Sub- and supercritical fluid extraction of functional ingredients from different natural sources: Plants, food-by-products, algae and microalgae: A review. *Food Chem.* **2006**, *98*, 136–148. [CrossRef]

19. Gilbert-López, B.; Mendiola, J.A.; Fontecha, J.; van den Broek, L.A.; Sijtsma, L.; Cifuentes, A.; Herrero, M.; Ibáñez, E. Downstream processing of *Isochrysis galbana*: A step towards microalgal biorefinery. *Green Chem.* **2015**, *17*, 4599–4609. [CrossRef]

20. Kim, S.M.; Kang, S.-W.; Kwon, O.-N.; Chung, D.; Pan, C.-H. Fucoxanthin as a major carotenoid in *Isochrysis aff. galbana*: Characterization of extraction for commercial application. *J. Korean Soc. Appl. Biol. Chem.* **2012**, *55*, 477–483. [CrossRef]

21. Ainsworth, E.A.; Gillespie, K.M. Estimation of total phenolic content and other oxidation substrates in plant tissues using Folin–Ciocalteu reagent. *Nat. Protoc.* **2007**, *2*, 875. [CrossRef]

22. Re, R.; Pellegrini, N.; Proteggente, A.; Pannala, A.; Yang, M.; Rice-Evans, C. Antioxidant activity applying an improved ABTS radical cation decolorization assay. *Free Radic. Biol. Med.* **1999**, *26*, 1231–1237. [CrossRef]

23. Del Pilar Sánchez-Camargo, A.; Montero, L.; Stiger-Pouvreau, V.; Tanniou, A.; Cifuentes, A.; Herrero, M.; Ibáñez, E. Considerations on the use of enzyme-assisted extraction in combination with pressurized liquids to recover bioactive compounds from algae. *Food Chem.* **2016**, *192*, 67–74. [CrossRef] [PubMed]

24. Lamers, P.P.; van de Laak, C.C.; Kaasenbrood, P.S.; Lorier, J.; Janssen, M.; De Vos, R.C.; Bino, R.J.; Wijffels, R.H. Carotenoid and fatty acid metabolism in light-stressed Dunaliella salina. *Biotechnol. Bioeng.* **2010**, *106*, 638–648. [CrossRef] [PubMed]

25. Conde, E.; Moure, A.; Domínguez, H. Supercritical CO_2 extraction of fatty acids, phenolics and fucoxanthin from freeze-dried Sargassum muticum. *J. Appl. Phycol.* **2015**, *27*, 957–964. [CrossRef]

26. Crampon, C.; Boutin, O.; Badens, E. Supercritical carbon dioxide extraction of molecules of interest from microalgae and seaweeds. *Ind. Eng. Chem. Res.* **2011**, *50*, 8941–8953. [CrossRef]

27. Kim, S.M.; Jung, Y.-J.; Kwon, O.-N.; Cha, K.H.; Um, B.-H.; Chung, D.; Pan, C.-H. A potential commercial source of fucoxanthin extracted from the microalga Phaeodactylum tricornutum. *Appl. Biochem. Biotechnol.* **2012**, *166*, 1843–1855. [CrossRef]

28. Xia, S.; Wang, K.; Wan, L.; Li, A.; Hu, Q.; Zhang, C. Production, characterization, and antioxidant activity of fucoxanthin from the marine diatom Odontella aurita. *Mar. Drugs* **2013**, *11*, 2667–2681. [CrossRef]

29. Jaswir, I.; Noviendri, D.; Salleh, H.M.; Taher, M.; Miyashita, K. Isolation of fucoxanthin and fatty acids analysis of Padina australis and cytotoxic effect of fucoxanthin on human lung cancer (H1299) cell lines. *Afr. J. Biotechnol.* **2011**, *10*, 18855–18862.

30. Kanazawa, K.; Ozaki, Y.; Hashimoto, T.; Das, S.K.; Matsushita, S.; Hirano, M.; Okada, T.; Komoto, A.; Mori, N.; Nakatsuka, M. Commercial-scale preparation of biofunctional fucoxanthin from waste parts of brown sea algae Laminalia japonica. *Food Sci. Technol. Res.* **2008**, *14*, 573. [CrossRef]

31. Fernández-Sevilla, J.M.; Fernández, F.A.; Grima, E.M. Biotechnological production of lutein and its applications. *Appl. Microbiol. Biotechnol.* **2010**, *86*, 27–40. [CrossRef]

32. Poojary, M.M.; Barba, F.J.; Aliakbarian, B.; Donsì, F.; Pataro, G.; Dias, D.A.; Juliano, P. Innovative alternative technologies to extract carotenoids from microalgae and seaweeds. *Mar. Drugs* **2016**, *14*, 214. [CrossRef]

33. Widowati, I.; Zainuri, M.; Kusumaningrum, H.P.; Susilowati, R.; Hardivillier, Y.; Leignel, V.; Bourgougnon, N.; Mouget, J.-L. Antioxidant activity of three microalgae Dunaliella salina, Tetraselmis chuii and *Isochrysis galbana* clone Tahiti. In Proceedings of the IOP Conference Series: Earth and Environmental Science, Bali, Indonesia, 25–27 October 2016; p. 012067.

34. Maadane, A.; Merghoub, N.; Ainane, T.; El Arroussi, H.; Benhima, R.; Amzazi, S.; Bakri, Y.; Wahby, I. Antioxidant activity of some Moroccan marine microalgae: Pufa profiles, carotenoids and phenolic content. *J. Biotechnol.* **2015**, *215*, 13–19. [CrossRef] [PubMed]

35. Foo, S.C.; Yusoff, F.M.; Ismail, M.; Basri, M.; Yau, S.K.; Khong, N.M.; Chan, K.W.; Ebrahimi, M. Antioxidant capacities of fucoxanthin-producing algae as influenced by their carotenoid and phenolic contents. *J. Biotechnol.* **2017**, *241*, 175–183. [CrossRef]

36. Michalak, I.; Dmytryk, A.; Wieczorek, P.P.; Rój, E.; Łęska, B.; Górka, B.; Messyasz, B.; Lipok, J.; Mikulewicz, M.; Wilk, R.; et al. Supercritical algal extracts: A source of biologically active compounds from nature. *J. Chem.* **2015**, *14*, 37.

37. Boonchum, W.; Peerapornpisal, Y.; Kanjanapothi, D.; Pekkoh, J.; Pumas, C.; Jamjai, U.; Amornlerdpison, D.; Noiraksar, T.; Vacharapiyasophon, P. Antioxidant activity of some seaweed from the Gulf of Thailand. *Int. J. Agric. Biol.* **2011**, *13*, 95–99.

38. Roh, M.-K.; Uddin, M.S.; Chun, B.-S. Extraction of fucoxanthin and polyphenol from Undaria pinnatifida using supercritical carbon dioxide with co-solvent. *Biotechnol. Bioprocess Eng.* **2008**, *13*, 724–729. [CrossRef]

39. Goiris, K.; Muylaert, K.; Fraeye, I.; Foubert, I.; De Brabanter, J.; De Cooman, L. Antioxidant potential of microalgae in relation to their phenolic and carotenoid content. *J. Appl. Phycol.* **2012**, *24*, 1477–1486. [CrossRef]

40. Li, H.-B.; Cheng, K.-W.; Wong, C.-C.; Fan, K.-W.; Chen, F.; Jiang, Y. Evaluation of antioxidant capacity and total phenolic content of different fractions of selected microalgae. *Food Chem.* **2007**, *102*, 771–776. [CrossRef]

41. Büyükokuroğlu, M.; Gülçin, I.; Oktay, M.; Küfrevioğlu, O. In vitro antioxidant properties of dantrolene sodium. *Pharmacol. Res.* **2001**, *44*, 491–494. [CrossRef]

42. Roginsky, V.; Lissi, E.A. Review of methods to determine chain-breaking antioxidant activity in food. *Food Chem.* **2005**, *92*, 235–254. [CrossRef]

43. Sachindra, N.M.; Sato, E.; Maeda, H.; Hosokawa, M.; Niwano, Y.; Kohno, M.; Miyashita, K. Radical scavenging and singlet oxygen quenching activity of marine carotenoid fucoxanthin and its metabolites. *J. Agric. Food Chem.* **2007**, *55*, 8516–8522. [CrossRef]

44. Reyes, F.A.; Mendiola, J.A.; Ibanez, E.; del Valle, J.M. Astaxanthin extraction from Haematococcus pluvialis using CO_2-expanded ethanol. *J. Supercrit. Fluids* **2014**, *92*, 75–83. [CrossRef]

45. Babova, O.; Occhipinti, A.; Capuzzo, A.; Maffei, M.E. Extraction of bilberry (Vaccinium myrtillus) antioxidants using supercritical/subcritical CO_2 and ethanol as co-solvent. *J. Supercri. Fluids* **2016**, *107*, 358–363. [CrossRef]

46. Kühn, S.; Temelli, F. Recovery of bioactive compounds from cranberry pomace using ternary mixtures of CO_2 + ethanol + water. *J. Supercrit. Fluids* **2017**, *130*, 147–155. [CrossRef]

47. Espinosa-Pardo, F.A.; Nakajima, V.M.; Macedo, G.A.; Macedo, J.A.; Martínez, J. Extraction of phenolic compounds from dry and fermented orange pomace using supercritical CO_2 and cosolvents. *Food Bioprod. Process.* **2017**, *101*, 1–10. [CrossRef]

48. Gallego, R.; Bueno, M.; Herrero, M. Sub- and supercritical fluid extraction of bioactive compounds from plants, food-by-products, seaweeds and microalgae—An update. *Trac-Trends Anal. Chem.* **2019**, *116*, 198–213. [CrossRef]

49. Di Mascio, P.; Murphy, M.E.; Sies, H. Antioxidant defense systems: The role of carotenoids, tocopherols, and thiols. *Am. J. Clin. Nutr.* **1991**, *53*, 194S–200S. [CrossRef] [PubMed]

50. Molino, A.; Larocca, V.; Di Sanzo, G.; Martino, M.; Casella, P.; Marino, T.; Karatza, D.; Musmarra, D. Extraction of Bioactive Compounds Using Supercritical Carbon Dioxide. *Molecules* **2019**, *24*, 782. [CrossRef] [PubMed]

51. Crampon, C.; Mouahid, A.; Toudji, S.-A.A.; Lépine, O.; Badens, E. Influence of pretreatment on supercritical CO_2 extraction from Nannochloropsis oculata. *J. Supercrit. Fluids* **2013**, *79*, 337–344. [CrossRef]

52. Wang, L.; Weller, C.L. Recent advances in extraction of nutraceuticals from plants. *Trends Food Sci. Technol.* **2006**, *17*, 300–312. [CrossRef]

53. Cikoš, A.-M.; Jokić, S.; Šubarić, D.; Jerković, I. Overview on the Application of Modern Methods for the Extraction of Bioactive Compounds from Marine Macroalgae. *Mar. Drugs* **2018**, *16*, 348. [CrossRef] [PubMed]

54. Machmudah, S.; Kawahito, Y.; Sasaki, M.; Goto, M. Supercritical CO_2 extraction of rosehip seed oil: Fatty acids composition and process optimization. *J. Supercrit. Fluids* **2007**, *41*, 421–428. [CrossRef]

55. Cheung, P.C. Temperature and pressure effects on supercritical carbon dioxide extraction of n-3 fatty acids from red seaweed. *Food Chem.* **1999**, *65*, 399–403. [CrossRef]

Article

Rhodotorula Strains Isolated from Seawater That Can Biotransform Raw Glycerol into Docosahexaenoic Acid (DHA) and Carotenoids for Animal Nutrition

Natalie L. Pino-Maureira [1,2,3,*], Rodrigo R. González-Saldía [1,2,3], Alejandro Capdeville [4] and Benjamín Srain [3]

1 Doctorado en Ciencias con Mención en Manejo de Recursos Acuáticos Renovables (MaReA), Facultad de Ciencias Naturales y Oceanográficas, Universidad de Concepción, Concepción 4070405, Chile; rogonzal@udec.cl
2 Departamento de Oceanografía, Facultad de Ciencias Naturales y Oceanográficas, Universidad de Concepción, Concepción 4070405, Chile
3 Centro de Investigación Oceanográfica COPAS Sur-Austral, Facultad de Ciencias Naturales y Oceanográficas, Universidad de Concepción, Concepción 4070405, Chile; bsrain@udec.cl
4 EWOS Chile, Alimentos Ltd.a., Parque Industrial, Escuadrón Km 20, Coronel 4190000, Chile; alejandro.capdeville.s@gmail.com
* Correspondence: napino@udec.cl; Tel.: +56-41-2983651

Featured Application: Marine biomass of *Rhodotorula* could be used as an alternative source of PUFAs and carotenoids for human and animal nutrition, using raw glycerol for their culture.

Abstract: Due to the overexploitation of industrial fisheries, as the principal source of fish oil, as well as the increasing replacement of synthetic pigments for animal nutrition, we need to find sustainable sources for these essential nutrient productions. Marine *Rhodotorula* strains NCYC4007 and NCYC1146 were used to determine the biosynthesis of docosahexaenoic acid (DHA) and carotenoids by biotransforming raw glycerol, a waste product of biodiesel. To evaluate the presence of inhibitory substances in raw glycerol, both strains were also grown in the presence of analytical grade glycerol and glucose as the main carbon source separately. With raw glycerol, NCYC4007 showed a significant correlation between DHA production and intracellular phosphorous concentrations. NCYC1146, a new *Rhodotorula* strain genetically described in this work, can produce canthaxanthin but only when glycerol is used as a main carbon source. Then, NCYC4007 could synthesize DHA as a phospholipid, and the production of canthaxanthin depends on the kind of carbon source used by NCYC1146. Finally, malate dehydrogenase activity and glucose production can be used as a proxy of the metabolisms in these marine *Rhodotorula*. This is the first evidence that marine *Rhodotorula* are capable of synthesizing DHA and canthaxanthin using an alternative and low-cost source of carbon to potentially scale their sustainable production for animal nutrition.

Keywords: *Rhodotorula* sp.; docosahexaenoic acid (DHA); carotenoids; canthaxanthin; raw glycerol

Citation: Pino-Maureira, N.L.; González-Saldía, R.R.; Capdeville, A.; Srain, B. Rhodotorula Strains Isolated from Seawater That Can Biotransform Raw Glycerol into Docosahexaenoic Acid (DHA) and Carotenoids for Animal Nutrition. *Appl. Sci.* **2021**, *11*, 2824. https://doi.org/10.3390/app11062824

Academic Editor: Marco F. L. Lemos

Received: 12 February 2021
Accepted: 5 March 2021
Published: 22 March 2021

Publisher's Note: MDPI stays neutral with regard to jurisdictional claims in published maps and institutional affiliations.

1. Introduction

Rhodotorula species are pigmented basidiomycetous yeasts in the family *Sporidiobolaceae* [1]. This genus contains 37 species, and only three of them, including *R. mucilaginosa* (formerly *R. rubra*), *R. minuta*, and *R. glutinis*, have been reported as causes of infection in humans [2]. Most species of *Rhodotorula* have been isolated from terrestrial ecosystems, which can metabolize diverse carbon sources. From the marine environment, the *Rhodotorula* described: *R. glutinis*, *R. glutinis* var. *dairiensis*, *R. aurantiaca*, *R. graminis*, *R. (mucilaginosa) rubra*, *R. pilimanae*, *R. minuta* and *R. aurea* are capable of metabolizing substrates such as galactose, lactose, maltose, sucrose, melibiose, raffinose, turanose, melezitose and cane molasses [3–5], and *Rhodotorula glutinis*, *R. mucilaginosa*, and *R. gracilis* are able to use glycerol and potato wastewater as a carbon source [6–8] but in low concentration of sodium ($\leq$1.1%).

From the marine ecosystem of the Southeastern Pacific twelve strains of marine fungoids capable of producing DHA, EPA and carotenoid pigments have been isolated [9]. In particular, the NCYC4007 yeast strain, characterized at the molecular level as a *Rhodotorula* strain, has been highlighted for its high production of DHA and carotenoids (23% and 11% dried weight, respectively) when the commercial Sabouraud medium (SM) was used [9]. This strain has been successfully used for fish larvae nutrition [10]. Additionally, in SM, a marine yeast from the Southern Austral marine ecosystem of Chile (strain NCYC1146), characterized as a high producer of carotenoids, was isolated. However, the genetic identity at the molecular level of this strain must be determined.

The strains NCYC4007 and NCYC1146 isolated from seawater seem to be promising candidates for the biotechnological production of DHA and carotenoid pigments. However, a crucial step to use them in that way needs to sort out the problem of the scaling the production using alternatives source of carbon with a lower price than commercial growing media (i.e., Sabouraud media). At the same time, the raising of the production of these substances to an industrial level needs to satisfy that higher economic, environmental and social sustainability can be reached.

In bioprocesses, raw glycerol is an important material for the economic viability of diverse fermentation processes. It is employed as alternative source of carbon different from glucose, due to the abundance and low cost. Among these are the production of substances of microbial origins such as 1, 3-propanediol, citric acid, succinic acid, polyhydroxyalkanoates and rhamnolipids by bacteria [11]; DHA and EPA by *Thraustochytrids* [12–18]; and lipids and carotenoid pigments by yeast isolated from terrestrial ecosystem [16–18]. Yeast isolated from seawater such as *Yarrowia lipolytica* has been cultivated using raw glycerol [19], although *R. marina*, a marine yeast member of the *Rhodotorula* genus, has not been cultivated using this raw substrate [17]. Other *Rhodotorula* members such as *R. glutinis* and *R. toruloides*, have been cultivated using raw glycerol [14,17]; however, they have not been isolated from seawater, and they require freshwater to grow.

In the present study, the capability of the marine *Rhodotorula* strain NCYC4007 and the marine yeast strain NCYC1146 to grow using raw glycerol in seawater and produce DHA and carotenoids was determined. To evaluate the presence of inhibitory substances in raw glycerol, both strains were grown in a medium with analytical grade glycerol and compared with the use of glucose as the main carbon source. The variables of the biomass production, DHA, total carotenoids and kinetic growth parameters were assessed. Subsequently, the fatty acid profile; carotenoid pigments; live microbial biomass and other metabolic parameters (glucose, phosphorus, cholesterol content and enzymatic activity of malate and lactate dehydrogenase) for both strains were characterized. Finally, the genetic characterization at the molecular level of the strain NCYC1146 was determined.

2. Materials and Methods

2.1. Molecular Genotyping of the NCYC 1146 Strain through 18s rRNA Gene Sequence

DNA extraction from the NCYC 1146 strain was performed using Ultra Clean® Microbial DNA Isolation (MO BIO Laboratories INC.) (Carlsbad, CA, USA) following the supplier's instructions. DNA was stored at -20 °C.

The amplification of the ribosomal gene 18s rRNA was carried out by PCR, using the primers described 18S1 (5′-AACCTGGTTGATCCTGCCAGTA-3′) and 18S12 (5′-CCITGTTACGACITCACCTTCCTCT- 3′) [18]. Reaction mixture contained the following reagents: 5× buffer of dNTPs, 2 mM of MgCl$_2$, 1 U of Go-taq polymerase, 0.5 µM of each primers and, approximately, 1.5 µL of DNA in a final volume of 25 µL. The amplification conditions used were specified by Honda et al. [18] as follows: initial denaturation at 95 °C for 5 min, followed by 35 amplification cycles. Each cycle consisted of denaturation at 95 °C for 30 s, alignment for 30 s at 55 °C and extension at 72 °C for 1 min, with a final extension step of 72 °C for 10 min. PCR products were visualized through 2% in agarose gel electrophoresis. The bands were visualized by gel staining with ethidium bromide (0.5 µg mL^{-1}) and exposure with a UV transilluminator.

Bands of expected sizes (>1500 pb) were purified using the E.Z.N.A® Gel Purification Kit, following the supplier's instructions. The purified PCR product was ligated to the pGEM®T-Easy vector (Promega) using a vector ratio of 3:1 in a final reaction volume of 10 µL and then incubated overnight at 4 °C.

Bacterial transformation was done with a vial of competent *Escherichia coli* cells (JM109® Promega) and the ligation product (recombinant plasmid), following the thermal shock methodology. The transformed bacteria were seeded in Petri dishes with LB agar and ampicillin (50 µg mL^{-1}), IPTG (0.5 mM) and β-galactosidase (50 mg mL^{-1}), which allowed for identifying the bacteria in the recombinant plasmid. The seeded plates were left to grow overnight at 37 °C. The positive clones were confirmed by conventional PCR in the colonies, and the bands were viewed by SafeView^{TM} staining with and exposure in a UV transilluminator. The positive colonies were grown in liquid LB medium with ampicillin (50 µg mL^{-1}), and the DNA plasmid was purified with the commercial kit commercial Plasmid Miniprep kit II E.Z.N.A.®, following the supplier's instructions.

Purified plasmids containing the cloned segment were sequenced by MACROGEN (South Korea). To identify the microorganisms, the sequences obtained were compared with the sequences available at the GenBank Database by the BLAST program available at the NCBI website. The phylogenetic tree based on the ribosomal gene 18s rRNA sequence was processed with the MEGA 6 and Geneious version 6.0.3 programs (Biomatters, Ltd.)

The sequences obtained from the studied strains were entered into the GenBank database with the following access numbers: KJ530974 (C6), KJ530975 (16CC1B), KJ530976 (NCYC 4007), KJ530977 (C3), KJ530978 (C30), KJ530979 (C46), KJ530980 (C51), KJ530981 (C24), KJ530982 (P39) and KJ530983 (C4).

2.2. Inoculum Production of NCYC 4007 and NCYC 1146

Two strains of marine yeast, NCYC 4007 and NCYC 1146, were used. Both strains isolated from the marine ecosystem of Chile (NCYC 4007 former C36 in [9]) were stored in the National Collection of Yeast Cultures (NCYC) in the United Kingdom under the Budapest Treaty. Strains were activated through the development of 100 mL of inoculum under Sabouraud growth medium at 21 °C for 7 days. Then, 500-mL flasks were used to cultivate an inoculum (10^6 cell mL^{-1}) for subsequent use in the evaluation of the growth media.

2.3. Raw Glycerol as a Carbon Source

In the present study, we evaluated as a carbon source: (1) raw glycerol originating from the biodiesel industry (RG), (2) analytical Grade glycerol (AG) and (3) glucose from commercial Sabouraud culture medium (SM). The RGM and AGM media were formulated with a 10% (*v/v*) of raw glycerol (nominal concentration) and 10% (*v/v*) of analytical grade of glycerol, respectively. Additionally, both mediums contained 7 g L^{-1} of peptone, 7 g L^{-1} of yeast extract, and 30 U mL^{-1} of penicillin–streptomycin. All media were performed with seawater (salinity = 35), filtered to 0.22 µm and autoclaved at 121 °C for 15 min. With the aim of selecting the best growth medium, these media were assessed at three temperatures (15, 21 and 31 °C), and the production of the biomass, DHA or carotenoids (NCYC 4007 and NCYC 1146, respectively) in a bioreactor (10 L) was evaluated.

In this study, the raw glycerol used corresponded to a highly concentrate fraction of glycerol collected during the industrial production of biodiesel from fried food waste oils (Comercial Verdemar Ltd. Company, COVEMAR-CHILE) (Talcahuano, Chile).

2.4. Production of Docosahexaenoic Acid (DHA) and Total Carotenoids (TC)

Aliquots taken every 24 h from the bioprocesses mentioned above were used to determine the amount of DHA and TC. Determination of DHA was carried out in three stages: (1) saponification extraction, (2) solvent changes and (3) chromatographic analysis, according to [20]. The saponification reaction was performed using 100 mg of wet biomass adding 1 mL of NaOH at 0.5 M in 96% ethanol. The cellular rupture was performed using an Ultra Turrax for 1 min. After this, it was centrifuged at 7000 rpm for 5 min, which dismissed any solid residue. One milliliter of HC1 was added to the obtained supernatant at 0.6 N and 3 mL of ethyl acetate placed in a vortex mixer for 1 min. The samples were incubated for 30 min at room temperature; to obtain phase separation, the upper phase was dried with N_2 (g) at room temperature. After this, the phases were freeze-dried to eliminate water residue. The freeze-dried samples were diluted in 400 µL of filtered methanol with polyethersulfone membranes RF-Jet Syringe Filter of 13 mm, with 0.45-µm pores and, after this, they were stored at $-20\,°C$ for analysis. For chromatographic analysis, a HPLC VWR™ HITACHI was used, a column LC-18 Supelco® of 15 cm $\times$ 4.6 mm. The mobile phase was performed at a gradient of 100% of A (75% of ACN LiChrosolv®) with a flow of 1 mL min^{-1} at 50% of B (pure ACN LiChrosolv®) with a flow of 2 mL min^{-1} over 15 min; consecutively, the flow changed to 1 mL min^{-1} until the 30-min cycle finished. Both A and B solutions were acidified with 0.12% of acetic acid (Sigma-Aldrich®, St. Louis, MO, USA). The detection was performed at 195 nm with an injection volume of 10 µL per sample. The DHA concentration was determined by a calibration curve performed using standard docosahexaenoic acid (Sigma-Aldrich®).

The determination of total carotenoids (TC) was performed by a methodology outlined by Rodher [20]. For this, 50 mg of wet biomass were weighed, obtained from aliquots during the bioprocesses. Once the samples were incubated and were centrifuged at 11,000 rpm for 10 min, the reached supernatant was transferred to a new tube for a further spectrophotometric analysis. Prior to the analysis, a potassium dichromate calibration curve was built of a stock solution of 3.6 mg mL^{-1}. Both the dilutions used to build the calibration curve and the samples were measured at 560 nm in glass cells. The TC concentration was determined, considering that 0.036% of potassium dichromate was equivalent to 2.6 ng mL^{-1} of TC.

2.5. Kinetic Growth and Biomass Production

The description of kinetic growth from the cultured strains was determined through a cell count every 12 h by a Neubauer Improved chamber. The obtained data were used to determine the kinetic parameters: maximum growth speed (µ maximum) and duplication time (Td). The total biomass produced was quantified at the end of each bioprocess, collected in 50-mL tubes, by spinning at 3600 rpm for 30 min and washed twice in distilled water. Following this, the biomass was freeze-dried and quantified in dry weight.

The quantification in dry weight was performed by two methods: (1) the quantification of the biomass being oven-dried at $100\,°C$ (oven-dried weight; ODW) and (2) quantification of the biomass by lyophilization (freeze-dried weight; FDW). In the case of the evaluation of the biomass obtained from the culture media, the percentages of DHA and TC were expressed as ODW. For the profiles of fatty acids and carotenoids, the biomass obtained corresponded to the FDW.

2.6. Fatty Acid Profiles

The extraction of lipids from the lyophilized cultures of *Rhodotorula* sp. NCYC 4007 was carried out according to a modified Bligh and Dyer [20], a procedure that substituted dichloromethane for chloroform. Samples (ca. 20 mg) were sequentially extracted by ultra-sonication with 30-mL dichloromethane/methanol (1:3 *v/v*, 2X, 1:1 *v/v*, 1X and dichloromethane (2X)). The lipid extract was concentrated with a rotary evaporator and dried with anhydrous Na_2SO_4.

Extracts were saponified with 15 mL of 0.5-N KOH: MeOH [20], and the non-saponifi able lipids were separated with hexane. The remaining aqueous extracts were acidified with 6-N HCl, and the fatty acids were extracted with hexane. The solvent was reduced by rotary evaporation, and the fatty acids were converted into methyl esters (FAMEs) with 1-mL 10% BF3/MeOH for 1 h at $70\,°C$ [21,22], 1 mL of milli-Q water was added and FAMEs was extracted with hexane. The organic fraction was then dried under a stream of N_2. Samples were injected into a gas chromatograph (Agilent 6890 GC series) coupled to a mass spectrometer detector (Agilent 5972 MS series) equipped with a HP5-MS column (30 m $\times$ 160 0.25 mm, 0.25-μm film thickness: Agilent Technologies) using He as a carrier gas. The oven temperature was 120 °C (2 min) to 290 °C at 4 °C min^{-1}, held for 5 min. The detector was operated in electron impact mode (70 eV) with an ion source at 230 °C. Mass spectra was acquired in full-scan mode (m/z range 40–600, scan rate 2.6 s^{-1}), and FAMEs were assigned using the retention times of a standard mixture (FAME mix, Supelco Analytical), the internal library of the mass spectrometer and the electronic data base available at website GoDaddy for details (GoDaddy, LLC). Quantification was carried out using a calibration curve with serial dilution of the FAME standard mix. The coefficient of variation for the analytical method was 14%, routinely measured with five replicate analyses.

2.7. Carotenoids Profiles

Biomass of the marine yeast strain NCYC 1146 was lyophilized and was used to determine the carotenoid profiles. For this, 1 g of freeze-dried biomass was weighed, and pigment extraction was performed with 50 mL of acetone. Extracts were centrifuged at 2000 rpm for 20 min, and 2 mL of acetone was added and centrifuged for 20 min. To dissolve oil droplets formed in the extracts (to avoid interference during reading), 1 mL of acetone was added to the supernatant. The samples were quantified in a HPLC Hitachi L6200, with a column RP-18 Supelco of 12.5-cm-diameter and 5-μm-sized particles. A mobile phase gradient system of water/ethyl acetate/methanol was used. A ratio of 2:10:88 (*v/v/v*) was used for 5 min at 0.75 mL min^{-1} for equilibration; flow was maintained for a further 10 min. Between 10 and 30 min, the solvent ratio was ramped to 2:50:48, and the flow rate was adjusted to 1.5 mL min^{-1}. A six-carotenoid standard, comprised of beta-carotene, canthaxanthin, astaxanthin, adonirrubin, zeaxanthin and 3-hidroxy echinenone, was used for identification purposes. These standard carotenoids were chosen, as they constitute the main starting, intermediate and final products within the microbial carotenoid pathway and have perceived commercial values.

2.8. Determination of Glucose, Phosphorus, Cholesterol and Live Microbial Biomass (ATP)

Aliquots of 1 mL were taken from the bioprocesses tested every 24 h for the subsequent glucose (intracellular and the culture media), phosphorus, cholesterol and live microbial biomass quantifications (ATP). The identification of glucose, phosphorus and intracellular cholesterol was performed through an extraction with phosphate buffer at 200 mM in ratio of 1:10 (p/v), homogenized with an Ultra Turrax for 1 min and centrifuged at 3000 rpm for 5 min. The supernatant obtained was utilized for the quantification. In the case of quantification phosphorus, it was determined the concentration of the buffer phosphate, and this value was used to normalize the samples. In the case of identifying glucose present in the culture liquid media, the samples were directly analyzed from the aliquots without prior treatment.

The quantification of glucose, phosphorus and cholesterol were performed using commercial kits by the company Human Diagnostics (Wiesbaden, Germany), following the supplier's instructions (Glucose liquicolor by the Glucose Oxidase (GOD)-Peroxidase (POD) method at 500 nm, Phosphorus liquirapid by phosphomolybdic acid at 340 nm and Cholesterol liquicolor by Cholesterol Oxidase (CHOD)–Peroxidase 4 Amino antipyrine (PAP) method at 500 nm).

To determine the fraction of ATP present in the aliquots, two steps were carried out according to the methodology described by Holm-Hansen and Booth [23]. The first one consisted of an ATP extraction through a boiling bath with organic buffer (Tris 20 mM, pH 7.7), and the second step, consistent with the ATP quantification extracted during the testing of bioluminescence for which an ATP meter, utilized Turner Designs Model TD 20/20. In addition, a calibration curve was performed between 0.39 nM and 100 nM from a stock solution of 1 M of ATP; through serial dilutions and to quantify the extracted ATP, FLE-250 was used to generate the complex enzyme substrate.

2.9. Enzymatic Activity of Malate Dehydrogenase (MDH) and Lactate Dehydrogenase (LDH)

The determination of enzymatic activity of MDH and LDH from the strains, aliquots (1 mL) were taken every 24 h and centrifuged at 11,000 rpm for 1 min. The supernatant obtained was extracted in a phosphate buffer 200 mM with an Ultra Turrax for 30 s and then centrifuged at 3000 rpm for 5 min. The enzymatic activities were quantified from the supernatants following the method described by González and Quiñones [24], and the enzymatic activities are shown in nmol of Nicotinamide Adenine Dinucleotide (NADH–reduced form) min^{-1} mL^{-1}. The activity of L-malate dehydrogenase was assayed, as it catalyzed the formation of malate from oxaloacetate. The assay medium contained 80-mM K_2HPO_4 buffer, pH 7.9, at 20 °C, 0.1-mM NADH, 150-mM $MgCl_2 * 6H_2O$ and 0.2-mM oxaloacetate. Absorption was monitored at 340 nm following the addition of the supernatant. Activity of LDH was measured with a reaction mixture that contained 80-mM K_2HPO_4 buffer, pH 7.9, at 20 °C, 3.2-mM pyruvate and 0.1-mM NADH. Absorbance was monitored at 340 nm following the addition of the supernatant. MDH and LDH activity measurements were corrected for nonspecific NADH oxidation.

2.10. Data Analysis

Statistical analysis was carried out on the results obtained from each of the tested cultures with the strains NCYC 4007 and NCYC 1146. In the first step, the homogeneity of variance (Bartlett, London, UK) and normality of data (Kolmogorov-Smirnov test) were analyzed. Consequently, the significance between DHA productivity, TC and total biomass was determined by one-way variance analysis (ANOVA). Furthermore, the analyses determined if significant correlations existed between the cell counts, enzymatic activities of MDH, LDH and live microbial biomass. All these analyses were carried out using the statistic software R Studio 3.3.1 (R. RStudio, PBC, Boston, MA, USA).

3. Results

3.1. Molecular Genotyping of Marine Yeast NCYC1146 Strain

The results obtained show that the amplification of the ribosome gene 18s rRNA from the NCYC 1146 strain generates a fragment of 1500 pb (data not shown). The analyses carried out by BLAST show that this strain shares a 97% similarity to species within the genus *Rhodotorula*. The strain is more distant to the species from genus *Rhodosporidium* and *Thraustochytrium* (91% and 81%, respectively). The phylogenetic tree shows the presence of two large clusters, one of which consisted of marine fungoid species, the producers of carotenoid pigments described by Pino et al. [9]. Another one was formed of NCYC4007, NCYC1146 strain and species from the genus *Rhodotorula* and *Rhodosporidium* (Figure 1).

3.2. Biomass Production and Optimum Growth Temperature of the Marine Rhodotorula sp. NCYC4007 and NCYC1146

Using raw and analytical grade glycerol, as well as glucose, as main source of carbon in a bioreactor of 10 L, the performance of the total produced biomass (% of FDW) at the end of the bioprocess (240 h) fluctuated between 54 and 152 g L^{-1} for the NCYC 4007 strain and between 49 and 192 g L^{-1} for NCYC1146 (Table 1). Nonetheless, the production of the biomass obtained from the culture of RGM at 15 °C is significantly higher ($p < 0.05$) for the NCYC4007 strain and that of NCYC1146 (Table 1). Consistently, the identified growth parameters described that the maximum growth speed (μ maximum) and the duplication time (Td) were also higher than the RGM at 15 °C for both strains (Table 1). Due to this, raw glycerol medium produces a significantly higher biomass in less time than other tested culture media at 15 °C ($p < 0.05$; Table 1), and there are no inhibitory effects on the growth.

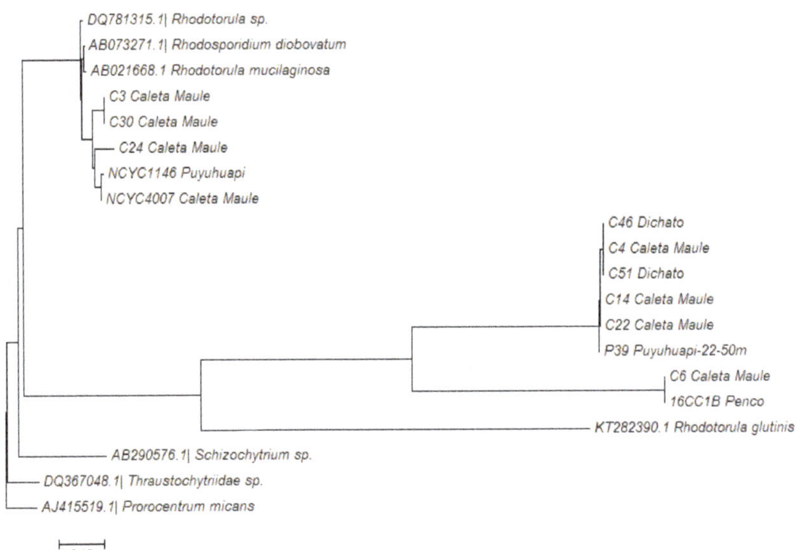

Figure 1. Location the NCYC1146 strain on the phylogenetic tree, based on the ribosomal gene 18s rRNA sequence, compared to the marine fungoides sequences described by Pino et al. [9]. To carry out the comparison, the species *Rhodotorula* sp., *Rhodotorula mucilaginosa, Rhodotorula glutinis, Rhodosporidium diobovatum, Schizochytrium* sp. and *Thraustochytriidae* sp. were used as representatives for the genus *Rhodotorula, Rhodosporidium, Schizochytrium* and *Thraustochytrium*. *Prorocentrum micans* was used as an outgroup. The tree was constructed by the Neighbor-joining (NJ) method and the MEGA 6 program.

Table 1. Kinetic growth parameters of the NCYC 4007 and NCYC 1146 strains obtained by the tested culture media at three different temperatures (15, 21 and 31 °C). Total biomass production in dry weight (B), maximum growth speed (μ maximum) and duplication time (Td).

Culture Media	B (g)		μ Maximum (h^{-1})		Td (h)	
	NCYC4007	NCYC1146	NCYC4007	NCYC1146	NCYC4007	NCYC1146
SM (15 °C)	108	97.2	0.03	0.05	23	14
SM (21 °C)	102	91.8	0.03	0.04	23	17
SM (31 °C)	96	86.4	0.03	0.04	23	17
AGM (15 °C)	126	113.4	0.03	0.05	23	14
AGM (21 °C)	116	104.4	0.03	0.04	23	17
AGM (31 °C)	126	113.4	0.05	0.07	14	10
RGM (15 °C)	*152	*191.8	*0.08	*0.11	*9	*6
RGM (21 °C)	92	82.8	0.04	0.06	17	12
RGM (31 °C)	54	48.6	0.05	0.07	14	10

* Significant statics value (<0.05).

3.3. DHA Production and Fatty Acid Profiles of NCYC4007 Strain

In raw and analytical grade glycerol, as well as glucose, the *Rhodotorula* strain NCYC 4007 presents a DHA production between 19% and 23% in ODW (Figure 2a–c). Comparing these three carbon sources, raw glycerol allowed the maximum DHA production at 144 h of cultivation, with a lesser time registered with analytical glycerol and glucose (192 and 216 h, respectively; Figure 2a,b). The results also indicated, at 15 °C, a significantly higher production of DHA (23% ODW; $p < 0.05$) when NCYC4007 was cultivated in raw glycerol (Figure 2c) compared to analytical grade glycerol and glucose was obtained. Thus, the DHA production in this strain is consistent from what was observed with respect to the biomass production in all carbon sources tested (Table 1). Given that this strain showed the greatest total DHA production (ODW) at 15 °C with the three carbon sources used, this was set as the optimum temperature to cultivate and acquire the fatty acid profiles of NCYC

4007 (Table 2). In general terms, the three carbon sources tested showed similar results for this strain, with a greater percentage of unsaturated fatty acids (UFAs) than saturated (SFAs). The high percentage of UFAs was achieved using glucose, followed by AG and RG (Table 2). The fatty acid profiles also showed that the NCYC 4007 strain produces a greater DHA percentage in respect to TFA using glucose (27% of TFA) than in the AG, albeit the DHA concentration (mg g^{-1} FDW biomass) obtained with glucose presented no significant differences with RG (Table 2).

3.4. Production of Total Carotenoids and Carotenoid Pigment Profiles of NCYC1146 Strain

For each of the three carbon sources used at the three tested temperatures, the TC production in the strain varied between 10% and 19% of the dry biomass (ODW; Figure 2d–f), and between 192 and 216 h, the maximum production was presented. With respect to the three tested media, the NCYC1146 strain presents a significantly greater production ($p < 0.05$) of TC with RG at 15 °C (19% ODW), like what was observed when this carbon source was used for DHA production with the NCYC4007 strain. The NCYC1146 strain produced the maximum total of TC at 192 h of cultivation (Figure 2f) in less time registered when glucose and AG was used (216 h; Figure 2d,e).

In the carotenoid profile derived from the biomass produced from the NCYC1146 strain, the presence of four pigments was observed (Figure 3). On average, the greatest quantity of substances produced were β-carotene (86.4 ± 20.5 µg g^{-1} FDW), followed by adonirrubin (48.8 ± 8.7 µg g^{-1} FDW), 3-hydroxyechinenone (44.2 ± 23.2 µg g^{-1} FDW) and canthaxanthin (24.8 ± 21.6 µg g^{-1} FDW). With glycerol (AG & RG), canthaxanthin a non-synthesized substance when glucose was produced. The concentration of this substance acquired from the final culture using both raw and analytical glycerol presented no significant differences (Figure 3).

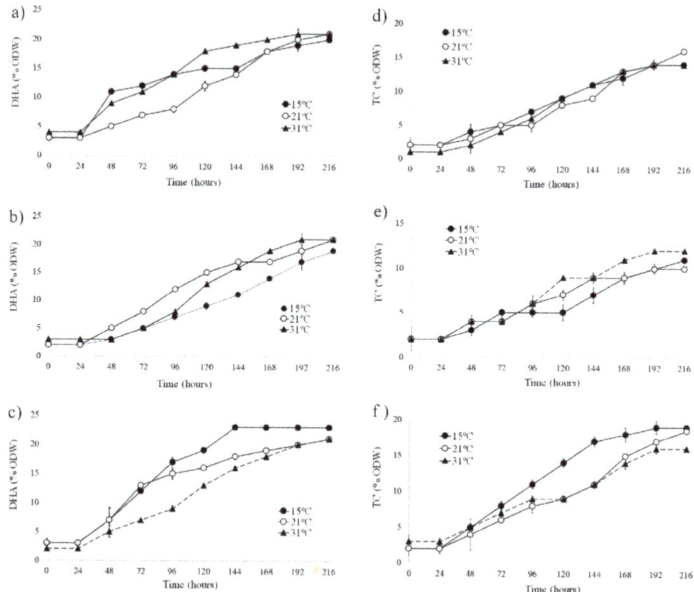

Figure 2. Average productions ± SD docosahexaenoic acid (DHA) and total carotenoids (TC) of the NCYC4007 and NCYC1146 strains. (**a**–**c**). Culture from the NCYC4007 strain in bioprocesses with SM, AGM and RGM media, respectively. (**d**–**f**). Culture from the NCYC1146 strain in bioprocesses with SM, AGM and RGM media, respectively. The production of these substances is shown in percentages, in respect to the biomass in oven-dried weight (ODW).

Table 2. Fatty acids profile from the NCYC 4007 strain cultivated in three different culture media at 15 °C: Sabouraud medium (SM), media formulated with analytic grade glycerol (AGM) and raw glycerol (RGM). The results are shown as the percentage of total fatty acid (TFA). In addition, the amount of fatty acids (FA) and DHA are showed in mg g^{-1} freeze-dried weight (FDW) biomass.

Carbon Source	SM	AGM	RGM
Fatty Acids	(%) TFA		
C14:0	0.2	0.2	0.8
C15:0	0.2	0.2	0.6
C16:0	3.8	4.5	8.9
C17:0	0.6	0.5	1.1
C18:0	0.0	4.7	10.5
C20:0	0.1	0.1	0.3
C22:0	0.2	0.2	0.5
C23:0	0.0	0.0	0.2
C24:0	0.6	0.5	1.0
C25:0	0.0	0.0	0.2
C26:0	0.0	0.0	0.1
C16:1 (ω9)	1.5	0.7	1.2
C17:1 (ω7)	1.0	0.3	0.7
C18:1 (ω9)	53.0	55.4	0.0
C18:2 (ω6)	4.6	3.0	42.2
C20:1 (ω9)	0.1	0.1	0.2
C20:5 (ω3)	7.2	10.2	9.2
C22:6 (ω3)	27.0	19.2	22.3
Total % SFA	5.6	11.2	24.1
Total % PUFA	94.4	88.9	75.8
Total FA (mg g^{-1})	46.2	45.3	53.6
Total DHA (mg g^{-1})	12.5	6.8	11.9

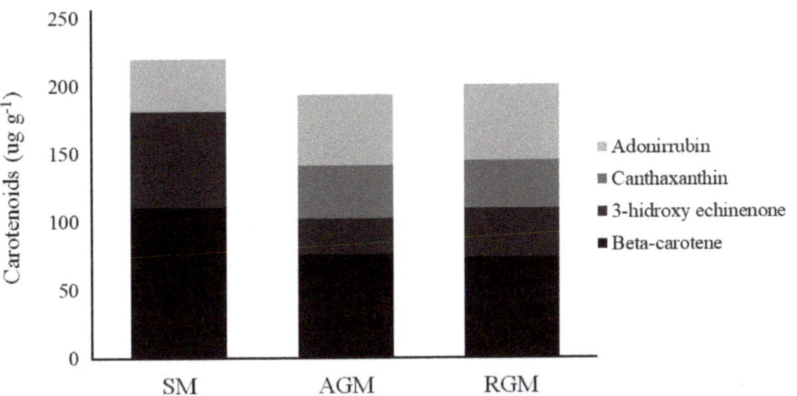

Figure 3. Carotenoid pigment profile of the NCYC1146 strain acquired from the total biomass produced from each of the bioprocesses carried out in the culture media SM, AGM and RGM at 15 °C. The results are shown in µg g^{-1}.

3.5. Metabolic Parameters of the Strains in Cultivated in Raw Glycerol

Using raw glycerol, the NCYC4007 strain at 15 °C presents a maximum production of DHA (23% DHA in ODW) at 144 h of cultivation and could maintain this production for four days (Figure 4a). A similar situation was observed for the NCYC1146 strain but with respect to carotenoid production (Figure 4b). For each strain, the live microbial biomass (ATP) was kept at basal levels until the culture reached the maximum DHA or TC production; from here on, an increase of ATP was observed and, subsequently, an increase in the cell count, which reached the maximum at 216 and 240 h for the NCYC4007 and NCYC1146 strains, respectively. (Figure 4a,b). In addition, observations were made at 24 h when the culture began the synthesis of carotenoids in the NCYC4007 strain, reaching values of 15% of CT in ODW at the end of the culture (Figure 4a). The intracellular glucose concentration determined in both strains presented a decrease in the time, contrary to what was observed in the culture medium, where an accumulation was observed during the cultivation (Figure 4c,d).

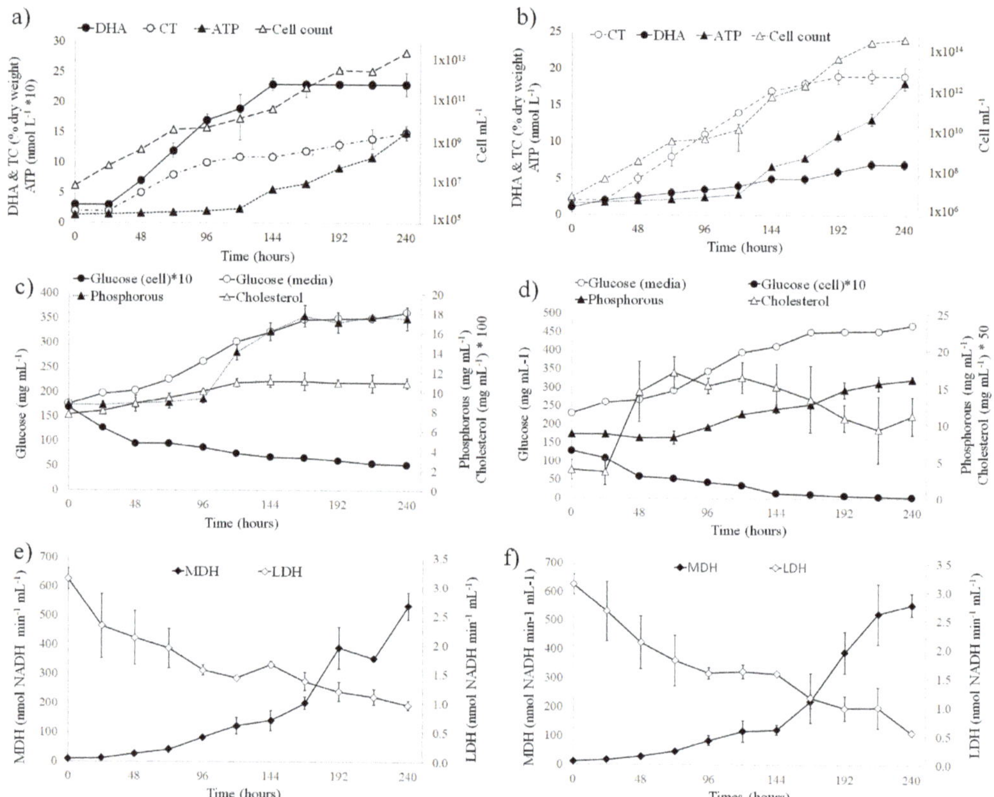

Figure 4. Average production ± SD of metabolic parameters determined for the NCYC4007 and NCYC1146 strains, cultivated in RGM at 15 °C. (**a,b**) Docosahexaenoic acid production (DHA), total carotenoids (TC), live microbial biomass (ATP) and cell count of strains NCYC4007 and NCYC1146, respectively. (**c,d**) Intracellular glucose concentration and in the culture media, total phosphorus and intracellular cholesterol of strains NCYC4007 and NCYC1146, respectively. (**e,f**) Enzymatic activities of MDH and LDH of the strains NCYC4007 and NCYC1146, respectively.

4. Discussion

The phylogenetic analysis of the ribosome gene 18s rRNA from the NCYC1146 strain was made with a fragment of similar size to that described by Honda et al. [19]. The NCYC1146 strain could be considered as a species of marine *Rhodotorula* sp., owing to the genetic similarity with the NCYC4007 strain (Figure 1) described by Pino et al. [9]. Despite the genetic similarity between NCYC4007 and NCYC1146, both strains can produce different substances that can be associated with the functional variability of the microorganisms when confronted by different environmental conditions. In this case, NCYC4007 comes from a temperate ecosystem with strong winds during upwelling periods [25], while

NCYC1146 is significantly different, isolated in a cold ecosystem with a high level of fresh water associated with adjacent rivers coming from the melting of snowdrifts and the high level of lignocellulosic material from a forestry origin [26].

This is the first report of marine *Rhodotorula* strains that can grow using raw glycerol as a carbon source. No inhibitory effects compared with analytical grade glycerol were observed. Using raw glycerol, the concordance observed between the significantly higher production of biomass and growth parameters at 15 °C (optimum growth temperature) for the marine *Rhodotorula* sp. NCYC4007 and NCYC1146 could be the result of a greater glycerol bioconversion to biomass compared to glucose. Non-marine *Rhodotorula* (*R. glutinis*, *R. mucilaginosa* and *R. gracilis*) showed that glycerol can be used to produce simultaneously lipids and carotenoids; nevertheless, a glycerol concentration higher than 10% can significatively decrease the growth rate [6,7]. *R. gracilis* seems to be a promising candidate for lipids and carotenoids production using low-cost wastes [8], with similar yields to the present study, but at a low sodium concentration. In the present study, the marine origin of the strains used can be an advantage to cope with the osmotic stress compared with non-marine *Rhodotorula*. In other marine microorganisms studied to produce lipids, such as *Thraustochytrium* sp. AMCQS5-5, it can produce a biomass eight times greater in the medium that contains glycerol instead of glucose as a carbon source [12]. In addition, in marine yeasts such as *Yarrowia lipolytica* has been described that the glycerol presents a higher absorption rate than the glucose, because glycerol passes into the microbial cell by facilitated diffusion and is assimilated via either the phosphorylation or the oxidative pathway [16,27].

The production of DHA registered in strain NCYC4007 (23% in ODW) is similar to those registered for *Thraustochytrium* sp. (20.4% in ODW; [28]), *Schizochytrium* KH105 (22% in ODW [29]) and *Schizochytrium limacinum* SR21 (19.5 at 22% in ODW [30]). In addition to these three tested media, the maximum production of DHA was observed between 144 and 216 h of cultivation, a greater time as described (96 to 120 h) for the *Thraustochytrium* sp., *Schizochytrium* KH105 and *Schizochytrium limacinum* SR21 strains.

In all carbon sources tested, a higher percent of UFAs than SFAs was observed in the fatty acids profile. Nevertheless, when raw glycerol is used, the lowest percentage of UFAs was observed. The change in the percentage of UFAs and SFAs as a result of the carbon source used has been reported in other oleaginous microorganisms such as *Rhodosporidium* sp. This strain observed a decrease in the percentage of PUFAs in respect to SFAs when glucose was used as a carbon source as opposed to glycerol [15]. Therefore, the raw glycerol used in this case could produce the same output of DHA as that obtained using glucose as a main carbon source from the commercial Sabouraud medium but using a substrate waste from another productive activity. This result is similar to that obtained in other marine fungoides such as *Thraustochytrium* sp. [31] and *Aurantiochytrium* sp. [32], where the total production of polyunsaturated fatty acids omega-3 do not demonstrate significant differences between glucose and raw glycerol.

A different way to compare the results from the carbon source used could be from a nutritional point of view. Accordingly, the relationship of unsaturated fatty acids in respect to saturated ones can be calculated for strain NCYC4007. In the study, the relationship indicates that the acquired biomass obtained when raw glycerol was used presents a lower degree of unsaturation (0.5), followed by analytical grade glycerol (0.9) and, finally, glucose (1.0). This indicator of unsaturation has been associated to the oxidative stability of fatty acids [33] and, in this case, suggests that lipid contents acquired from the biomass obtained when glucose was used have a greater possibility to oxidize than lipids produced when raw glycerol was used. Thus, the acquired biomass with raw glycerol, despite having a lower unsaturation level (0.5), could present a greater stability of the PUFAs against oxidation in a bioprocess on a larger productive purposes scale than that of the biomass acquired with glucose as a carbon source (1.0).

In greater detail, the acquired profiles obtained when glucose and analytical grade glycerol were used as a carbon source highlight the presence of oleic acid (C18:1; Table 2), a precursor in the synthesis of polyunsaturated fatty acids belonging to the omega-3 family, such as DHA and omega-6 [34]. Meanwhile, with raw glycerol, a high production of linoleic acid was observed (C18:2, Table 2), a precursor in the synthesis of fatty acids belonging to the omega-6 family, which can be transformed via catalysis Δ15 desaturase to α-linoleic acid (C18:3) and, also, is a precursor to DHA synthesis [34]. These results suggest that the NCYC4007 strain could have more than one metabolic pathway for DHA synthesis, which could be induced depending on which substrate is used, according to what was observed in *Schizochytrium* [28], *Aurantiochytrium* and *Thraustochytrium* [35].

The TC production observed in strain NCYC1146 (19% in ODW) is greater than described for *R. mucilaginosa* [13] and *R. glutinis* [14], corresponding to an 8% or 10% TC of ODW, respectively. The maximum production of TC in less time was observed when raw glycerol is used, similar to what was observed for the production of DHA in NCYC4007. This observation could be related with a greater activity of the biosynthesis routes due to the use of raw glycerol as a substrate as opposed to glucose. This has been described as a species of *Thraustochytrids*, where the polyketide synthase pathway (PKS) and that of mevalonate were involved in DHA production, used for the production of carotenoids and found to be more active in a glycerol medium to one with glucose [36].

The carotenoid pigments profile of strain NCYC1146 is similar to the profiles from *Thraustochytrids strains* AMCQS5-3, AMCQS5-5 [12] and ONC-T18 [36] that produce the greatest concentration of β-carotene, followed by echinenone (intermediary of 3-hydroxyechinenone) and canthaxanthin. It was observed that extending the culture time of the *Thraustochytrids* ONC-T18 strain [37] could synthesize other xanthines such as astaxanthin due to 3-hydroxyechinenone, and the adonirrubin are precursors of this substance [37], both present in the carotenoids profile of NCYC1146 (Figure 3). To increase the cultivation time of NCYC1146 will be addressed in the future to determine if this strain can produce astaxanthin in greater cultivation times than in this study.

The *Thraustochytrids* have demonstrated that the DHA and carotenoid production are related across acetyl-coenzyme A (CoA) and the availability of NADPH, both used in the synthesis of these substances [14]. Therefore, the carotenoid total production and DHA as secondary metabolites of the NCYC4007 and NCYC1146 strains, respectively, would indicate that the use of raw glycerol could provide a good supply of carbon for the synthesis of said substances. In this way, DHA is synthesized from malonyl-CoA coming from Acetyl-CoA, which, at the same time, is a key precursor for the synthesis of acetoacetyl-CoA used for the subsequent biosynthesis of carotenoids and squalene in *Thraustochytrids* [14]. In this way, the simultaneous synthesis of both DHA and carotenoids in these marine *Rhodotorula*, using raw glycerol as a substrate, suggests that this microorganism should possess the analog biosynthetic pathways described by *Thraustochytrids*. Subsequent studies should clarify whether both strains also have the ability to synthesize squalene, a substance that shares the biosynthetic pathways described above and that also has commercial value, as well as DHA and carotenoids.

The carotenoid concentration with raw glycerol reached by strain NCYC 4007 is greater than observed by Pino et al. [9] for the same strain but using glucose as a carbon source. The NCYC1146 strain also synthesizes the DHA (7% ODW of DHA at the end of the study; Figure 4b), substance of which in preliminary studies just reached 0.3% ODW when the strain was cultured with glucose. The synthesis of a secondary metabolite was observed in the cultures of the NCYC4007 (carotenoids) and NCYC1146 strains (DHA) when raw glycerol was used as a substrate, which suggests that an interconnection should exist between the biosynthesis of these two substances in these marine *Rhodotorula*, as observed in *Thraustochytrium* sp. AMCQS5-5 [12] and *Thraustochytrids* species [14].

To the best of our knowledge, the glucose accumulation in the culture medium when raw glycerol was used as a carbon source has not been reported in oleaginous microorganisms. Nevertheless, the appearance when raw glycerol was used could be that the excess glycerol promotes a gluconeogenesis and subsequently eliminates the synthesized glucose from the cell. To determine glucose in the culture medium could be a useful parameter to determine the optimum raw glycerol concentration that could be used. It has been observed that an excess of glycerol can produce an inhibition of DHA synthesis in *Schizochytrium limacinum* [38] and *Schizochytrium limacinum* SR21 [30] when used in percentages greater than 8.5% and 9%, respectively, values inferior to those used in this study (10%). In addition, excess glycerol has demonstrated the inhibition of β-carotene synthesis in *Blakeslea trispora* [39] and, furthermore, could restrict xanthine production, which uses β-carotene as a precursor [14].

Another interesting result to consider is the sudden increase in the intracellular phosphorus concentration from 96 h in NCYC4007, which is not as noticeable in NCYC1146 (Figure 4c,d). In NCYC4007, the intracellular phosphorus concentrations present a significant and strong positive correlation (semi-log; $R^2 = 0.8217$; $p = 0.001195$) with the DHA concentrations; what is suggested is that this latter could be synthesized as a phospholipid rather than as a triacylglycerol. If the DHA synthesized by NCYC4007 were in the phospholipid form, this would have a greater oxidative stability, as well as intestinal bioavailability, if used for human and animal nutrition [40]. The DHA synthesis in NCYC1146 is minor to that of NCYC4007; nonetheless, the concentrations of this substance in NCYC1146 also correlate significantly with the intracellular phosphorus of this strain (semi-log; $R^2 = 0.8999$; $p = 8.589 \times 10^{-6}$).

The intracellular cholesterol concentration in NCYC4007 (Figure 4c) presents a positive and significant relation with the intracellular phosphorus concentrations (semi-log; $R^2 = 0.7835$; $p = 0.00092$). In addition, the intracellular cholesterol presents a significant relation with the DHA concentrations ($R^2 = 0.961$; $p = 1.207 \times 10^{-7}$), so this is another precedent in favor that synthesized DHA of this strain could be a phospholipid. It has been reported that cholesterol presents a high interaction with phospholipids when determined in marine species [41]. No relationship was observed between cholesterol and phosphorus in NCYC 1146; however, the main product in this strain is not DHA but carotenoids.

In respect to enzymatic activities, the increase in the MDH activity was observed in both strains and a decrease in the LDH activity during the course of the cultures (Figure 4e,f). The decrease of LDH activity could be associated to the use of pyruvate (anaerobic pathway) for the synthesis of the first fatty acid precursor of DHA (palmitic acid; C16:0) [42]. In addition, the MDH activity for both strains presented a strong correlation of log–log interactions with the cell count (NCYC 4007, $R^2 = 0.9596$, $p = 0.044$ and NCYC 1146, $R^2 = 0.9566$, $p = 0.000002$) and live microbial biomass (NCYC 4007, $R^2 = 0.8566$, P= 0.00004 and NCYC 1146, $R^2 = 0.8578$, $p = 0.00001$). These results indicate that MDH activity can be used as a proxy of the metabolic activity for both strains, similar to those described in the facultative anaerobic microorganisms [43] and microplankton organisms in the ocean [23].

5. Conclusions

The DHA and TC production by NCYC4007 and NCYC 1146, respectively, is greater in RGM at 15 °C. Both strains used raw glycerol as a common precursor for biosynthesis but at different rates. DHA synthetized by NCYC4007 could be a phospholipid, and NCYC 1146 could be synthetized as a canthaxanthin. The metabolic parameters suggest that the glucose accumulation in the medium and intracellular MDH activity could be used as a good proxy to adjust the optimum glycerol concentration. Finally, these strains could be used as an alternative source of PUFAs and carotenoids for human and animal nutrition, using raw glycerol for their culture.

6. Patents

In this work are two patent applications obtained entered into the Instituto Nacional de Propiedad Industrial (INAPI), Chile: (1) Flaked food for ornamental fish and larval stages of fish, S 1224. (2) A pelleted food rich in polyunsaturated fatty acids (PUFAs) for salmonid fingerlings, S 3471. Additionally, there was an International Patent Application through the Patent Cooperation Treaty (PCT), titled: A pelleted food rich in polyunsaturated fatty acids (PUFAs) for salmonid fingerlings, PCT/CL2018/050158.

Author Contributions: N.L.P.-M. and R.R.G.-S. designed the experiments and analyzed the data; N.L.P.-M. performed the experiments, validation to scale the bioprocess in the bioreactor 10 L and statistically analyzed method; A.C. provided the analysis of the samples in HPLC for the profile of the carotenoids; B.S. provided the analysis of the samples in the chromatography of gases (GC-MM) for the profiles of the fatty acids and N.L.P.-M. and R.R.G.-S. wrote the paper. All authors have read and agreed to the published version of the manuscript.

Funding: This study could be carried out thanks to financial assistance from CONICYT ANID 21161591, CONICYT PIA PFB31 and APOYO CCTE AFB170006, Universidad de Concepción, Concepcion, Chile.

Acknowledgments: The authors would like to thank the Comercial Verdemar Ltd. Company (COVEMAR), a producer of biodiesel that facilitated raw glycerol, considered by them to be a residue. Finally, the help of Marine Organic Geochemistry Laboratory and EWOS Chile Alimentos Ltd.a., is appreciated for their collaboration in the fatty acid and carotenoid profile determinations, respectively.

Conflicts of Interest: The authors declare no conflict of interest.

References

1. Fell, J.W.; Boekhout, T.; Fonseca, A.; Scorzetti, G.; Statzell-Tallman, A. Biodiversity and systematics of basidiomycetous yeasts as determined by large-subunit rDNA D1/D2 domain sequence analysis. *Int. J. Syst. Evol. Microbiol.* **2000**, *50*, 1351–1371. [CrossRef]
2. Biswas, S.K.; Yokoyama, K.; Nishimura, K.; Miyaji, M. Molecular phylogenetics of the genus *Rhodotorula* and related basidiomycetous yeasts inferred from the mitochondrial cytochrome b gene. *Int. J. Syst. Evol. Microbiol.* **2001**, *51*, 1191–1199. [CrossRef]
3. Ahearn, D.G.; Roth, F.J., Jr.; Meyers, S.P. A comparative study of marine and terrestrial strains of *Rhodotorula*. *Can. J. Microbiol.* **1962**, *8*, 121–132. [CrossRef]
4. Jiru, T.M.; Steyn, L.; Pohl, C.; Abate, D. Production of single cell oil from cane molasses by *Rhodotorula kratochvilovae* (syn, *Rhodosporidium kratochvilovae*) SY89 as a biodiesel feedstock. *Chem. Cent. J.* **2018**, *12*, 1–7. [CrossRef]
5. Prabhu, A.A.; Gadela, R.; Bharali, B.; Deshavath, N.N.; Dasu, V.V. Development of high biomass and lipid yielding medium for newly isolated *Rhodotorula mucilaginosa*. *Fuel* **2019**, *239*, 874–885. [CrossRef]
6. Kot, A.M.; Błażejak, S.; Kieliszek, M.; Gientka, I.; Bryś, J. Simultaneous production of lipids and carotenoids by the red yeast *Rhodotorula* from waste glycerol fraction and potato wastewater. *Appl. Biochem. Biotechnol.* **2019**, *189*, 589–607. [CrossRef]
7. Kot, A.M.; Błażejak, S.; Kieliszek, M.; Gientka, I.; Bryś, J.; Reczek, L.; Pobiega, K. Effect of exogenous stress factors on the biosynthesis of carotenoids and lipids by *Rhodotorula* yeast strains in media containing agro-industrial waste. *World J. Microbiol. Biotechnol.* **2019**, *35*, 1–10. [CrossRef] [PubMed]
8. Kot, A.M.; Błażejak, S.; Kieliszek, M.; Gientka, I.; Piwowarek, K.; Brzezińska, R. Production of lipids and carotenoids by *Rhodotorula gracilis* ATCC 10788 yeast in a bioreactor using low-cost wastes. *Biocatal. Agric. Biotechnol.* **2020**, *26*, 101634. [CrossRef]
9. Pino, N.L.; Socias, C.; González, R.R. Marine fungoid producers of DHA, EPA and carotenoids from central and southern Chilean marine ecosystems. *Rev. Biol. Mar. Oceanogr.* **2015**, *50*, 507–520. [CrossRef]
10. Barra, M.; Llanos-Rivera, A.; Cruzat, F.; Pino-Maureira, N.; González-Saldía, R. The marine fungi *Rhodotorula* sp. (Strain CNYC4007) as a potential feed source for fish larvae Nutrition. *Mar. Drugs* **2017**, *15*, 369. [CrossRef] [PubMed]
11. Papanikolaou, S.; Aggelis, G. Modelling aspects of the biotechnological valorization of raw glycerol: Production of citric acid by *Yarrowia lipolytica* and 1, 3- propanediol by *Clostridium butyricum*. *J. Chem. Technol. Biotechnol.* **2003**, *78*, 542–547. [CrossRef]
12. Gupta, A.; Singh, D.; Barrow, C.J.; Puri, M. Exploring potential use of Australian thraustochytrids for the bioconversion of glycerol to omega-3 and carotenoids production. *Biochem. Eng. J.* **2013**, *78*, 11–17. [CrossRef]
13. Aksu, Z.; Eren, A.T. Carotenoids production by the yeast *Rhodotorula mucilaginosa*: Use of agricultural wastes as a carbon source. *Process Biochem.* **2005**, *40*, 2985–2991. [CrossRef]
14. Saenge, C.; Cheirsilp, B.; Suksaroge, T.T.; Bourtoom, T. Potential use of oleaginous red yeast *Rhodotorula glutinis* for the bioconversion of crude glycerol from biodiesel plant to lipids and carotenoids. *Process Biochem.* **2011**, *46*, 210–218. [CrossRef]
15. Xu, J.; Zhao, X.; Wang, W.; Du, W.; Liu, D. Microbial conversion of biodiesel byproduct glycerol to triacylglycerols by oleaginous yeast *Rhodosporidium toruloides* and the individual effect of some impurities on lipid production. *Biochem. Eng. J.* **2012**, *65*, 30–36. [CrossRef]
16. Makri, A.; Fakas, S.; Aggelis, G. Metabolic activities of biotechnological interest in *Yarrowia lipolytica* grown on glycerol in repeated batch cultures. *Bioresour. Technol.* **2010**, *101*, 2351–2358. [CrossRef]
17. Mączka, W.; Wińska, K.; Grabarczyk, M.; Żarowska, B. Biotransformation of α-Acetylbutyrolactone in *Rhodotorula* Strains. *Int. J. Mol. Sci.* **2018**, *19*, 2106. [CrossRef] [PubMed]
18. Honda, D.; Yokochi, T.; Nakahara, T.; Raghukumar, S.; Nakagiri, A.; Schaumann, K.; Higashihara, T. Molecular phylogeny of *labyrinthulids* and *thraustochytrids* based on the sequencing of 18S ribosomal RNA gene. *J. Eukaryot. Microbiol.* **1999**, *46*, 637–647. [CrossRef] [PubMed]

19. Rodher, A. Determinación de carotenoids. *Rev. Agroquím. Tecnol. Aliment.* **1966**, *6*, 24–27.
20. Bligh, E.G.; Dyer, W.J. A rapid method for total lipid extraction and purification. *Can. J. Biochem. Physiol.* **1959**, *37*, 911–917. [CrossRef]
21. Christie, W.W. The preparation of derivate of fatty acids. In *Gas Chromatography and Lipids: A Practical Guide*; Christie, W.W., Ed.; The Oily Press: Bridgwater, UK, 1989; pp. 36–39.
22. Tolosa, I.; Vescovali, I.; Leblond, N.; Marty, J.C.; de Mora, S.; Prieur, L. Distribution of pigments and fatty acids biomarkers in particulate matter from the frontal structure of the Alboran Sea (SW Mediterranean Sea). *Mar. Chem.* **2004**, *88*, 103–125. [CrossRef]
23. Holm-Hansen, O.; Both, C. The measurement of adenosine triphosphate in the ocean and its ecological significance. *Limnol. Oceanogr.* **1966**, *11*, 510–519. [CrossRef]
24. González, R.R.; Quiñones, R.A. Common catabolic enzyme patterns in a microplankton community of the Humboldt Current System off northern and central-south Chile: Malate dehydrogenase activity as an index of water-column metabolism in an oxygen minimum zone. *Deep-Sea Res. Part II Top. Stud. Oceanogr.* **2009**, *56*, 1095–1104. [CrossRef]
25. Schneider, W.; Donoso, D.; Garcés-Vargas, J.; Escribano, R. Water-column cooling and sea surface salinity increase in the upwelling region off central-south Chile driven by a poleward displacement of the South Pacific High. *Prog. Oceanogr.* **2017**, *151*, 38–48. [CrossRef]
26. Pérez-Santos, I.; Garcés-Vargas, J.; Schneider, W.; Ross, L.; Parra, S.; Valle-Levinson, A. Double-diffusive layering and mixing in Patagonian fjords. *Prog. Oceanogr.* **2014**, *129*, 35–49. [CrossRef]
27. Morgunov, I.G.; Kamzolova, S.V.; Lunina, J.N. The citric acid production from raw glycerol by *Yarrowia lipolytica* yeast and its regulation. *Appl. Microbiol. Biotechnol.* **2013**, *97*, 7387–7397. [CrossRef] [PubMed]
28. Metz, J.G.; Roessler, P.; Facciotti, D.; Levering, C.; Dittrich, F.; Lassner, M.; Yazawa, K. Production of polyunsaturated fatty acids by polyketide synthases in both prokaryotes and eukaryotes. *Science* **2001**, *293*, 290–293. [CrossRef] [PubMed]
29. Yamasaki, T.; Aki, T.; Mori, Y.; Yamamoto, T.; Shinozaki, M.; Kawamoto, S.; Ono, K. Nutritional enrichment of larval fish feed with *thraustochytrid* producing polyunsaturated fatty acids and xanthophylls. *J. Biosci. Bioeng.* **2007**, *104*, 200–206. [CrossRef] [PubMed]
30. Chi, Z.; Pyle, D.; Wen, Z.; Frear, C.; Chen, S. A laboratory study of producing docosahexaenoic acid from biodiesel-waste glycerol by microalgal fermentation. *Process Biochem.* **2007**, *42*, 1537–1545. [CrossRef]
31. Avalos, J.; Nordzieke, S.; Parra, O.; Pardo-Medina, J.; Limón, M.C. Carotenoid production by filamentous fungi and yeasts. In *Biotechnology of Yeasts and Filamentous Fungi*; Sibirny, A.A., Ed.; Springer: Berlin, Germany, 2017; pp. 225–279.
32. Chang, K.J.L.; Dumsday, G.; Nichols, P.D.; Dunstan, G.A.; Blackburn, S.I.; Koutoulis, A. High cell density cultivation of a novel *Aurantiochytrium* sp. strain TC 20 in a fed-batch system using glycerol to produce feedstock for biodiesel and omega-3 oils. *Appl. Microbiol. Biotechnol.* **2013**, *97*, 6907–6918. [CrossRef]
33. Hernandez, A.G.D. Nutrition treatise: Composition and nutritional quality of foods Tome II., Ed. *Med. Panam.* **2010**, *2*, 812.
34. Béligon, V.; Christophe, G.; Fontanille, P.; Larroche, C. Microbial lipids as potential source to food supplements. *Curr. Opin. Food Sci.* **2016**, *7*, 35–42. [CrossRef]
35. Nagano, N.; Sakaguchi, K.; Taoka, Y.; Okita, Y.; Honda, D.; Ito, M.; Hayashi, M. Detection of genes involved in fatty acid elongation and desaturation in thraustochytrid marine eukaryotes. *J. Oleo Sci.* **2011**, *60*, 475–481. [CrossRef]
36. Aasen, I.M.; Ertesvåg, H.; Heggeset, T.M.B.; Liu, B.; Brautaset, T.; Vadstein, O.; Ellingsen, T.E. Thraustochytrids as production organisms for docosahexaenoic acid (DHA), squalene, and carotenoids. *Appl. Microbiol. Biotechnol.* **2016**, *100*, 4309–4321. [CrossRef]
37. Armenta, R.E.; Burja, A.; Radianingtyas, H.; Barrow, C.J. Critical assessment of various techniques for the extraction of carotenoids and co-enzyme Q10 from the *Thraustochytrid* strain ONC-T18. *J. Agric. Food Chem.* **2006**, *54*, 9752–9758. [CrossRef]
38. Yokochi, T.; Honda, D.; Higashihara, T.; Nakahara, T. Optimization of docosahexaenoic acid production by Schizochytrium limacinum SR21. *Appl. Microbiol. Biotechnol.* **1998**, *49*, 72–76. [CrossRef]
39. Mantzouridou, F.; Naziri, E.; Tsimidou, M.Z. Industrial glycerol as a supplementary carbon source in the production of β-carotene by Blakeslea trispora. *J. Agric. Food Chem.* **2008**, *56*, 2668–2675. [CrossRef]
40. Henna, L.; Nielsen, N.; Timm-Heinrich, M.; Jacobsen, C. Oxidative stability of marine phospholipids in the liposomal form and their applications. *Lipids* **2011**, *46*, 3–23. [CrossRef] [PubMed]
41. Valenzuela, A.; Valenzuela, R.; Sanhueza, J.; de la Barra, F.; Morales, G. Fosfolípidos de origen marino: Una nueva alternativa para la suplementación con ácidos grasos omega-3. *Rev. Chil. Nutr.* **2014**, *41*, 433–438. [CrossRef]
42. Christophe, G.; Kumar, V.; Nouaille, R.; Gaudet, G.; Fontanille, P.; Pandey, A.; Soccol, C.R.; Larroche, C. Recent developments in microbial oils production: A possible alternative to vegetable oils for biodiesel without competition with human food? *Braz. Arch. Biol. Technol.* **2012**, *55*, 29–46. [CrossRef]
43. Fenchel, T.; Finlay, B.J. *Ecology and Evolution in Anoxic Worlds*; Oxford University Press: Oxford, UK, 1995; 276p.

Article

Industry-Friendly Hydroethanolic Extraction Protocols for *Grateloupia turuturu* UV-Shielding and Antioxidant Compounds

Rafael Félix , **Ana M. Carmona**, **Carina Félix**, **Sara C. Novais** and **Marco F. L. Lemos** *

MARE—Marine and Environmental Sciences Centre, ESTM, Instituto Politécnico de Leiria,
2520-641 Peniche, Portugal; rafael.felix@ipleiria.pt (R.F.); carmonamarta1@gmail.com (A.M.C.);
carina.r.felix@ipleiria.pt (C.F.); sara.novais@ipleiria.pt (S.C.N.)
* Correspondence: marco.lemos@ipleiria.pt

Received: 26 June 2020; Accepted: 29 July 2020; Published: 31 July 2020

Featured Application: To develop industry-friendly extraction methodologies to obtain bioactive extracts of *Grateloupia turuturu* for the cosmeceutical industry.

Abstract: *Grateloupia turuturu* is an invasive macroalga on the Iberian coast, known to produce bioactive compounds with different cosmeceutical bioactivities, namely UV shielding and antioxidants. The goal of this study was to optimize the extraction procedure of main bioactivities of this species with cosmetic potential, using Response Surface Methodology. Two Box–Behnken designs were used to evaluate the effect of ethanol concentration (0–50%), liquid-solid ratio, time, pH, and temperature on yield, UV absorbance, and antioxidant activity. Both optimizations showed a similar trend: aqueous extracts have higher yields and extracts performed with ethanol as part of the solvent have higher activities concerning UV absorbance and antioxidant activity. For all the extracts an absorption peak between 320 and 340 nm was observed. This data now allows further studies by narrowing the extracts worthful of characterization. The development of industry-friendly extraction methods allows the valorization of this invasive species, contributing for the potential creation of natural and eco-friendly products by the cosmetic industry while contributing to the restoration of affected environments.

Keywords: Box–Behnken design; extraction conditions; bioactive compounds; invasive seaweed; cosmetics

---

1. Introduction

The use of marine organisms for the exploitation of bioactive compounds has recently increased due to their biotechnological potential in different areas as pharmaceutical, cosmetic/cosmeceutical, food/feed, and many others [1,2]. Within this group, macroalgae are known to be great producers of bioactive compounds [2], with growth rates that ensure biomass availability from harvesting and aquaculture.

Grateloupia turuturu is an edible red macroalga native from Korea and Japan that has been classified as invasive species in the Atlantic Ocean, and was first reported in Portugal in 1997 [3]. This seaweed is characterized as rich in carbohydrates and proteins, as well as having a low lipid content [4]. It is present in "mid-intertidal" areas and is easily found in intertidal pools [5], where the exposure to solar radiation and oxidative stress levels are high [5]. The ability to protect themselves against harmful radiation make these organisms potential sources for the discovery of compounds with antioxidants and UV shielding activities [4,6]. Moreover, it has product development potential in the cosmetics industry.

For a successful screening of such compounds and to ensure the industrial feasibility of their application, an adequate method of extraction should be developed [7]. Although more recent and effective methods have been developed (the novel extraction methodologies, characterized by more technological approaches resulting in better performance and lower environmental impact), the utilization of solid-liquid extraction (SLE) is still one of the best solutions for the production of extracts with the intent of industrialization, given the low technical and facility requirements for its up-scale implementation. However, SLE can be fine-tuned to perform better (be more efficient, selective, eco-friendly, and profitable) by optimizing several operational parameters: physicochemical properties of the raw material, the selected solvents/respective concentrations, pH, temperature, extraction time, and others [8]. A balance between the time of extraction and the temperature is extremely important to avoid thermal degradation. Room temperature can be an advantage, as it preserves compounds and reduces the associated economic costs of the procedure [7]. Regarding the type of solvent, water-based extractions are privileged and ethanol and acetone are the most accepted organic solvents considering their security and cost [9]. Solvent selection depends not only on their inherent safety and cost but also on the cost associated to their evaporation, and ability to extract target compounds. For all these reasons, binary mixtures of water and ethanol are popular in the natural extract production for industry. Specifically, in the case of *G. turuturu*, hydroethanolic extraction is the most biotechnologically relevant since this is the best solvent to recover this species' main bioactivities: sulphated galactans (known antioxidant, anticoagulant, and antimicrobial activities [10]), phycobiliproteins (PBPs), and mycosporine-like amino acids (MAAs) (with antioxidant activity and very high UV-shielding activity [4,10]).

The optimization of the extraction parameters for the obtention of compounds groups described above using a one-dimensional method, where only one factor is modified at a time, is not only time-consuming but more expensive [11]. Using the response surface methodology (RSM), it is possible to avoid these difficulties, since this statistical method allows for the simultaneous evaluation of different variables (enabling the detection of potential interactions between the tested conditions), reducing the high number of experiments, and, consequently, the quantity of reagents used and the time associated to the process [12].

Thus, the main goal of this study was to optimize the extraction process of bioactivities as measured by antioxidants and UV-shielding from the red macroalga *Grateloupia turuturu*, with the potential for use in the cosmetic industry. Reponses surface methodology was used to characterize the effects of ethanol percentage in the extraction solvent, time of extraction, liquid-to-solid ratio, temperature, and pH in these bioactivities. This promoted the added value to the species through the potential creation of natural and eco-friendly products in the cosmetics field, while contributing for the restoration of the natural environments, following a circular economy approach.

2. Materials and Methods

2.1. Seaweed Collection

Grateloupia turuturu was collected at Aguda beach in Arcozelo, Portugal (41.054826, −8.656865). The collected biomass was cleaned and sorted for epibionts, dried in a wind tunnel at 25 °C, milled to flour-like powder (particle size 150 ± 50 μm), and vacuum stored at room temperature, in the dark, until further use.

2.2. Response Surface Methodology: Box–Behnken Design

An RSM with a Box–Behnken design was performed to determine the influence of three independent parameters in the hydro-ethanolic extracts: liquid-solid ratio (mL·g^{-1}) (LSR), time of extraction (min), and ethanol percentage, as established below (Table 1).

Table 1. Independent variables tested during the first optimization of the extraction process using a Box–Behnken design: ethanol percentage (X_1), liquid-solid ratio (LSR) (X_2), and extraction time (X_3).

Run	Coded Variables			Actual Variable Values		
	(X_1)	(X_2)	(X_3)	% EtOH (*v/v*) (X_1)	LSR (mL·g^{-1}) (X_2)	Time (Minutes) (X_3)
1	−1	0	−1	0	25	20
2	1	0	−1	50	25	20
3	−1	0	1	0	25	100
4	1	0	1	50	25	100
5	−1	−1	0	0	10	60
6	−1	1	0	0	40	60
7	1	−1	0	50	10	60
8	1	1	0	50	40	60
9	0	−1	−1	25	10	20
10	0	−1	1	25	10	100
11	0	1	−1	25	40	20
12	0	1	1	25	40	100
13	0	0	0	25	25	60
14	0	0	0	25	25	60
15	0	0	0	25	25	60

A second assay was designed to assess the influence of the extraction temperature (°C), pH, and ethanol percentage (*v/v*) (Table 2). In this case, the LSR and extraction time were defined according to results obtained from the first optimization assay (time of extraction of 60 min and LSR of 40 mL·g^{-1}).

Table 2. Independent variables tested during the second optimization of the extraction process using a Box–Behnken design: ethanol percentage (X_1), extraction temperature (X_2), and pH (X_3).

Run	Coded Variables			Actual Variable Values		
	(X_1)	(X_2)	(X_3)	% EtOH (*v/v*) (X_1)	Temperature (°C) (X_2)	pH (X_3)
1	0	−1	−1	25	25	4
2	0	−1	1	25	25	10
3	0	1	−1	25	95	4
4	0	1	1	25	95	10
5	−1	0	−1	0	60	4
6	1	0	−1	50	60	4
7	−1	0	1	0	60	10
8	1	0	1	50	60	10
9	−1	−1	0	0	25	7
10	−1	1	0	0	95	7
11	1	−1	0	50	25	7
12	1	1	0	50	95	7
13	0	0	0	25	60	7
14	0	0	0	25	60	7
15	0	0	0	25	60	7

Extracts of *Grateloupia turuturu* were obtained by mixing 5.0 g of the seaweed biomass with the assigned volume of the assigned solvent, for the assigned time under constant magnetic stirring and thermostatized. After the extraction procedure, each extract was centrifuged (10,000× *g* for 5 min) and the supernatant filtered at low pressure (Whatmann n°1) and stored at 4 °C until further processing (maximum 24 h). The extracts were then evaporated under reduced pressure at 40 °C and desiccated under a vacuum concentrator at room temperature (Vacufuge, Eppendorf, Germany). Dry extracts were weighed and yield was calculated (g extract·g^{-1} biomass).

Each extract was then resuspended for further assaying their UV-absorbance and antioxidant activities. Water extracts were resuspended in water at 25 mg·mL^{-1}; 25% (*v/v*) ethanol extracts were resuspended in 25% (*v/v*) DMSO in water at 50 mg·mL^{-1}; 50% (*v/v*) ethanol extracts were resuspended in 50% (*v/v*) DMSO in water at 100 mg·mL^{-1}.

2.3. UV Absorbance Spectra

For the UV absorption spectrum, 200 µL of each extract at 0.1 mg·mL^{-1} was added to a 96-well microplate suitable for UV readings (n = 4; Greiner UV-Star®, Greiner Bio-one, Kremsmünster, Austria) as well as the respective blanks, following the methodology of Maciel et al. [13]. The absorbance was read between the wavelengths of 280 nm and 400 nm (Biotek® Synergy H1, Winooski, VT, USA). The integral of the absorbance (Abs) between these wavelengths was computed and used to calculate a massic extinction coefficient (ε, in mL·cm^{-1}·g^{-1} extract), using the liquid height in each well (l, in cm) and mass concentration of the extracts (C, in mg·mL^{-1}) for each extract according to the formula:

$$\varepsilon = \frac{\int_{280}^{400} Abs}{l * C}$$

2.4. Antioxidant Activity

The antioxidant activity was measured with the oxygen radical absorbance capacity (ORAC) assay, according to Dávalos and colleagues [14]. For the preparation of the standard curve, a Trolox stock solution, diluted in 75 mM phosphate buffer (VWR®, Radnor, PA, USA), pH 7.4, was used to prepare the dilutions from 8 µM to 0.5 µM. The obtained extracts were tested at 1 mg·mL^{-1} (diluted in 75 mM phosphate buffer). A fluorescein solution at 70 nM was used and the AAPH reagent (Sigma®, Darmstadt, Germany) was prepared fresh at a final concentration of 12 mM. A total of 20 µL of each sample were placed in a 96-well black microplate (Greiner®, Kremsmünster, Austria) and 120 µL of a fluorescein solution (70nM) (Sigma®, Darmstadt, Germany) was added to all samples, including the standard curve. For blanks, 120 µL of 75 mM phosphate buffer was added instead of fluorescein. The plate was then incubated inside the microplate reader (Biotek® Synergy H1, Winooski, VT, USA) at 37 °C and the fluorescence was read for 15 min with 1 min interval at an excitation wavelength of 485 nm, and an emission wavelength of 525 nm. After the incubation period, 60 µL of pre-warmed (37 °C) AAPH (2,2'-Azobis(2-methylpropionamidine) dihydrochloride) reagent (Sigma®, Darmstadt, Germany) was added. The fluorescence was read for 80 min with 1 min intervals. Data was treated and results were expressed as µmol of Trolox equivalents per gram of extract (µmol TE·g^{-1} ext).

2.5. Data Treatment

The values of antioxidant activity and of UV-absorbance were studied both as specific activity (activity per mass unit of extract) and total activity (activity per unit of seaweed extracted), the latter calculated by multiplying the values of the respective activities by the yield of extract. The results from the two Box–Behnken were organized in a table containing 30 experimental runs (the 30 extracts), for which the value of each of the five variables under study (percentage of ethanol, LSR, time, temperature, and pH), and the value for each of the five parameters determined (yield, specific and total antioxidant activity, and specific and total UV-absorbance) was available.

For each Box–Behnken experiment, five response surface models were computed by multiple linear regression. For that, R statistical language was used, along with RStudio, and packages 'rsm' and 'viridis' [15–18]. The basis for model development was a quadratic multiple linear regression containing all three variables, but no interactions, followed by a manual stepwise, backwards-forward-backwards regression fitting. Then, the model was manually refined by considering the *p*-value of the estimates β, the *p*-value of the model and both the R-squared and the adjusted R-squared of the model. Individual factors with a *p*-value greater than 0.1 were always excluded and individual factors with a *p*-value between 0.05 and 0.1 were excluded whenever their removal from the model resulted in an approximation of the adjusted R-squared to the R-squared. Then, linear binary interactions of the variables were added iteratively to sporadically improve the model's fit, because ethanol percentages introduced a chemical variability that could cause drastic shifts in the parameters, quadratic binary interactions, and cubic variables. We attempted to maintain the degrees of freedom above 8 and keep residue distribution random via visual inspection of the fit-real plot. The equations and goodness-of-fit statistical parameters (residual standard error, multiple R-squared, adjusted R-squared, *p*-value, and degrees of freedom) are reported in the manuscript alongside the contour plots for each of the five models at three slices of ethanol percentage. The fit-real plots (Figure S1), as well as the individual *p*-values of the estimates (Figure S2), and the complete dataset are included in the Supplementary Material.

3. Results

Response surface methodology was used to characterize the effects of LSR, time of extraction, and ethanol percentage in the solvent on the bioactivities recovered from *G. turuturu* biomass using a Box–Behnken design of experiment. Moreover, a second experiment using LSR and time values allowed the characterization of temperature and pH effects on different hydroethanolic extractions. Both optimizations showed a similar trend: aqueous extracts have higher yields and extracts performed with ethanol as part of the solvent have higher activities concerning the UV absorbance and antioxidant activity. For all the extracts, an absorption peak between 320 and 340 nm was observed.

3.1. First Optimization

The models of the effects of ethanol concentration (X_1), LSR (X_2), and time of extraction (X_3) are presented in Figures 1–5. Concerning the yield of these extracts (Figure 1), the aqueous extractions varied between 0.25 and 0.5 g extract·g^{-1} biomass with an almost linear increase with an increasing LSR for all times. For extracts with 50% ethanol (*v/v*), yield varied from 0.2 and 0.25 g extract·g^{-1} biomass with maximums at either central time and maximum LSR, or minimum LSR at either of the time extremes. For intermediate ethanol percentages, the yield varied between 0.2 and 0.38 g extract·g^{-1} biomass, increasing when LSR increased.

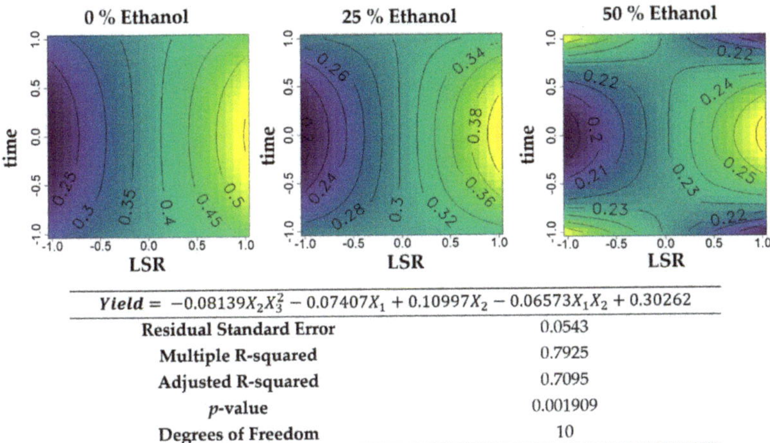

$$Yield = -0.08139X_2X_3^2 - 0.07407X_1 + 0.10997X_2 - 0.06573X_1X_2 + 0.30262$$

Residual Standard Error	0.0543
Multiple R-squared	0.7925
Adjusted R-squared	0.7095
p-value	0.001909
Degrees of Freedom	10

Figure 1. Contour plots of the modelled function regarding the yield obtained during the first optimization of *Grateloupia turuturu* hydroethanolic solid-liquid extraction (SLE) using a Box–Behnken design, in the presence of different concentrations of solvent (ethanol). For each plot, XX axis indicates the liquid-solid ratio (LSR) and YY axis indicates extraction time. Color gradients represents the increase (yellow color) and decrease (blue color) in yield according to the tested variables. Below, the equation and goodness-of-fit statistical parameters for the variable Yield (g extract·g^{-1} biomass) in the first optimization process, according to the ethanol concentration (X_1), LSR (X_2), and extraction time (X_3) are presented.

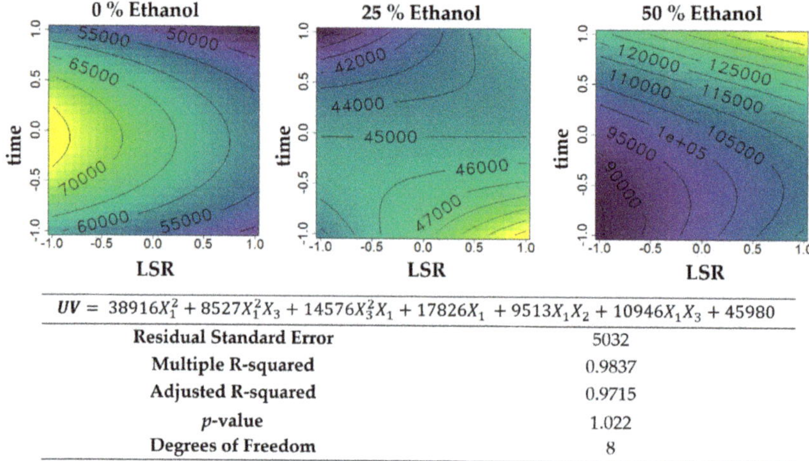

$$UV = 38916X_1^2 + 8527X_1^2X_3 + 14576X_3^2X_1 + 17826X_1 + 9513X_1X_2 + 10946X_1X_3 + 45980$$

Residual Standard Error	5032
Multiple R-squared	0.9837
Adjusted R-squared	0.9715
p-value	1.022
Degrees of Freedom	8

Figure 2. Contour plots of the modelled function regarding the UV absorbance (per gram of extract) obtained during the first optimization of *Grateloupia turuturu* hydroethanolic SLE using a Box–Behnken design, in the presence of different concentrations of solvent (ethanol). For each plot, XX axis indicates the LSR and YY axis indicates extraction time. Color gradients represents the increase (yellow color) and decrease (blue color) in UV absorbance according to the tested variables. Below, the equation and goodness-of-fit statistical parameters for the variable UV absorbance (mL·cm^{-1}·g^{-1} extract) in the first optimization process, according to the ethanol concentration (X_1), (X_2), and extraction time (X_3) are presented.

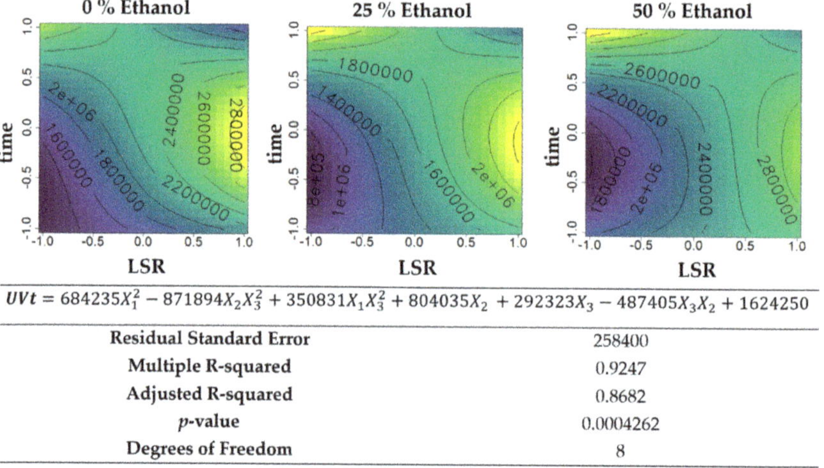

$$UVt = 684235X_1^2 - 871894X_2X_3^2 + 350831X_1X_3^2 + 804035X_2 + 292323X_3 - 487405X_3X_2 + 1624250$$

Residual Standard Error	258400
Multiple R-squared	0.9247
Adjusted R-squared	0.8682
p-value	0.0004262
Degrees of Freedom	8

Figure 3. Contour plots of the modelled function regarding the total UV absorbance (per gram of biomass) obtained during the first optimization of *Grateloupia turuturu* hydroethanolic SLE using a Box–Behnken design, in the presence of different concentrations of solvent (ethanol). For each plot, XX axis indicates the LSR and YY axis indicates extraction time. Color gradients represents the increase (yellow color) and decrease (blue color) in total UV absorbance according to the tested variables. Below, the equation and goodness-of-fit statistical parameters for the variable total UV absorbance (mL·cm^{-1}·g^{-1} biomass) in the first optimization process, according to the ethanol concentration (X_1), LSR (X_2), and extraction time (X_3) are presented.

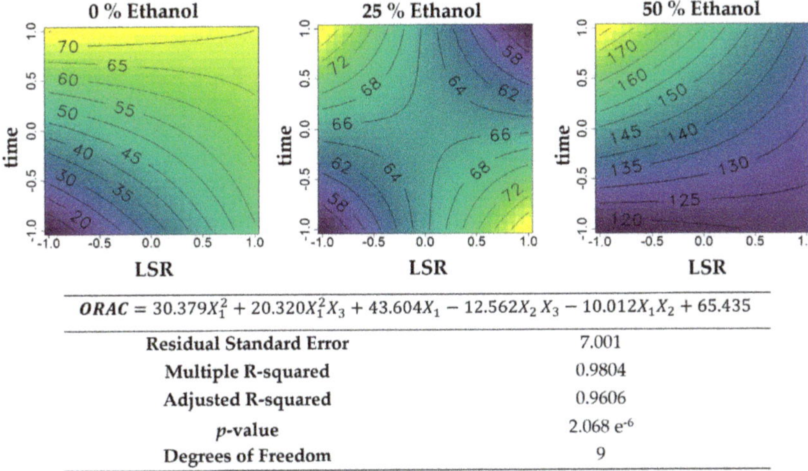

$$ORAC = 30.379X_1^2 + 20.320X_1^2X_3 + 43.604X_1 - 12.562X_2X_3 - 10.012X_1X_2 + 65.435$$

Residual Standard Error	7.001
Multiple R-squared	0.9804
Adjusted R-squared	0.9606
p-value	2.068 e^{-6}
Degrees of Freedom	9

Figure 4. Contour plots of the modelled function regarding the oxygen radical absorbance capacity (ORAC) activity (per gram of extract) obtained during the first optimization of *Grateloupia turuturu* hydroethanolic SLE using a Box–Behnken design, in the presence of different concentrations of solvent (ethanol). For each plot, XX axis indicates the LSR and YY axis indicates extraction time. Color gradients represents the increase (yellow color) and decrease (blue color) in ORAC activity according to the tested variables. Below, the equation and goodness-of-fit statistical parameters for the variable antioxidant activity by ORAC method (μmol TE·g^{-1} extract) in the first optimization process, according to the ethanol concentration (X_1), LSR (X_2), and extraction time (X_3) are presented.

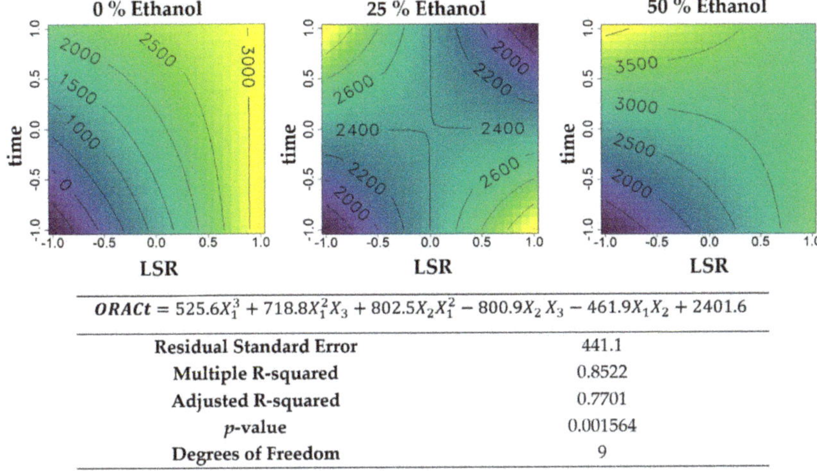

$$ORACt = 525.6X_1^3 + 718.8X_1^2X_3 + 802.5X_2X_1^2 - 800.9X_2X_3 - 461.9X_1X_2 + 2401.6$$

Residual Standard Error	441.1
Multiple R-squared	0.8522
Adjusted R-squared	0.7701
p-value	0.001564
Degrees of Freedom	9

Figure 5. Contour plots of the modelled function regarding the total ORAC activity (per gram of biomass) obtained during the first optimization of *Grateloupia turuturu* hydroethanolic SLE using a Box–Behnken design, in the presence of different concentrations of solvent (ethanol). For each plot, XX axis indicates the LSR and YY axis indicates extraction time. Color gradients represents the increase (yellow color) and decrease (blue color) in total ORAC activity according to the tested variables. Below, the equation and goodness-of-fit statistical parameters for the variable total ORAC activity (μmol TE·g^{-1} biomass) in the first optimization process, according to the ethanol concentration (X_1), LSR (X_2), and extraction time (X_3) are presented.

Regarding UV absorbance per gram of extract (Figure 2), for the aqueous extractions, it varied between 50,000 and 70,000 mL·cm^{-1}·g^{-1} extract, reaching its maximum with lower LSR and central time of extraction, decreasing with increasing LSR and extreme times of extraction. In the presence of 50% ethanol (*v/v*), higher UV absorbance values were obtained, increasing with higher times of extraction and LSR, while the intermediate percentage of ethanol showed higher values when high LSR and low time of extraction are combined. Globally, the results showed that increasing concentrations of ethanol favor the UV absorbance of extracts.

The total UV absorbance per gram of biomass (Figure 3), showed higher values in aqueous extractions and hydroethanolic extractions with 50% ethanol (*v/v*), reaching the maximum value of 2,800,000 mL·cm^{-1}·g^{-1} biomass. For aqueous extractions, the minimum was reached with low LSR and low time of extraction and maximum with higher LSR and a central time of extraction. The hydroethanolic extractions followed a similar tendency.

Considering the antioxidant activity per gram of extract (Figure 4), an increase was verified when ethanol percentages increased, varying from 20 to 70 µmol TE·g^{-1} extract in aqueous extractions, and between 120 and 170 µmol TE·g^{-1} extract in hydroethanolic extractions with 50% ethanol (*v/v*). LSR did not influence antioxidant activity in aqueous extractions when combined with higher time of extraction, where the maximum activity is recovered. On the contrary, in the presence of 50% ethanol (*v/v*), higher time of extraction and lower LSR generated the highest values. For intermediate percentages of ethanol, maximum values were found in the presence of higher time of extraction and minimum LSR, and with higher LSR and lower extraction time.

The total ORAC activity per gram of biomass (Figure 5) maximized hydroethanolic extractions with 50% of ethanol (*v/v*), followed by aqueous extractions. For water-based extractions, the maximum experimental value (3000 µmol TE·g^{-1} biomass) was reached in the presence of high LSR, independent of extraction time, showing a lower LSR and a lower time the minimum value found, while for 50% ethanol (*v/v*), the antioxidant activity is maximized in the presence of higher extraction times and lower LSR.

The overall analysis of the models led to the decision to fix the LSR at 40 mL·g^{-1} and time of extraction at 60 min in the second Box–Behnken design.

3.2. Second Optimization

From the second round of extractions that included the variables ethanol concentration (%) (X_1), temperature (°C) (X_2), and pH (X_3), the same models described above were attained (Figures 6–10). Globally, yield of cold extraction followed the same trend as before (lower values with more ethanol) and that of hot extraction was almost constant across ethanol concentrations. Yield (Figure 6) had maximum values in hydroethanolic extractions using 50% ethanol (*v/v*) for central temperatures, close to 60 °C and acidic pH. For aqueous extractions and extractions with 25% ethanol, the minimum obtained was 0.4 g extract·g^{-1} biomass and the maximum 0.6 g extract·g^{-1} biomass. However, a basic pH and high temperature were necessary to reach the maximum value in aqueous extractions, while for hydroethanolic extractions with 25% ethanol (*v/v*) the maximum was reached with lower temperatures and acidic pH.

Concerning the UV absorbance per gram of extract (Figure 7), an increase of the values with increasing percentages of ethanol was achieved, being the maximum value of aqueous extractions 38,000 mL·cm^{-1}·g^{-1} extract, for 25% ethanol (*v/v*) 40,000 mL·cm^{-1}·g^{-1} extract and 50% ethanol (*v/v*) 80,000 mL·cm^{-1}·g^{-1} extract. Using the temperature of 25 °C for water-based extractions, the maximum was reached regardless of pH. For the highest percentage of ethanol, the maximum values were attained with basic pH and lower temperatures.

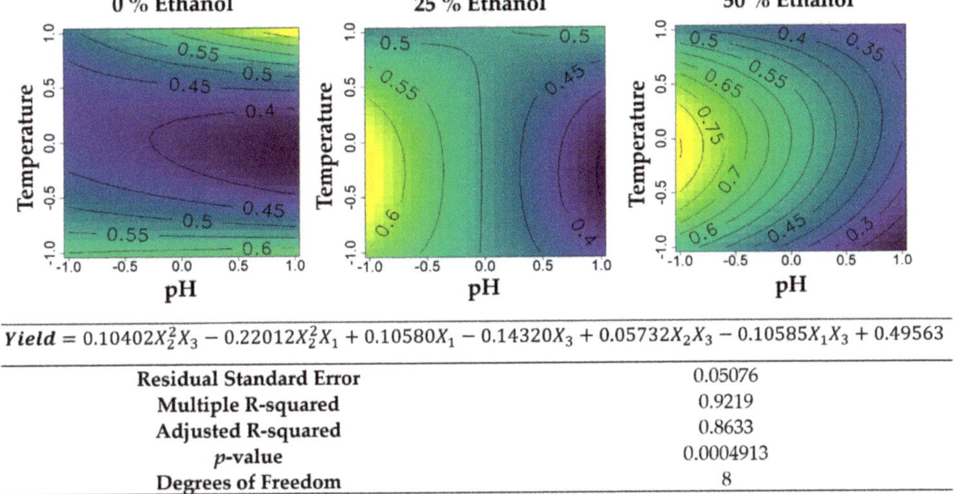

$$Yield = 0.10402X_2^2X_3 - 0.22012X_2^2X_1 + 0.10580X_1 - 0.14320X_3 + 0.05732X_2X_3 - 0.10585X_1X_3 + 0.49563$$

Residual Standard Error	0.05076
Multiple R-squared	0.9219
Adjusted R-squared	0.8633
***p*-value**	0.0004913
Degrees of Freedom	8

Figure 6. Contour plots of the modelled function regarding the yield obtained during the second optimization of *Grateloupia turuturu* hydroethanolic SLE using a Box–Behnken design, in the presence of different concentrations of solvent (ethanol). For each plot, XX axis indicates the pH and YY axis indicates temperature. Color gradients represents the increase (yellow color) and decrease (blue color) in yield according to the tested variables. Below, the equation and goodness-of-fit statistical parameters for the variable yield (g extract·g^{-1} biomass) in the second optimization process, according to the ethanol concentration (X_1), temperature (X_2), and pH (X_3) are presented.

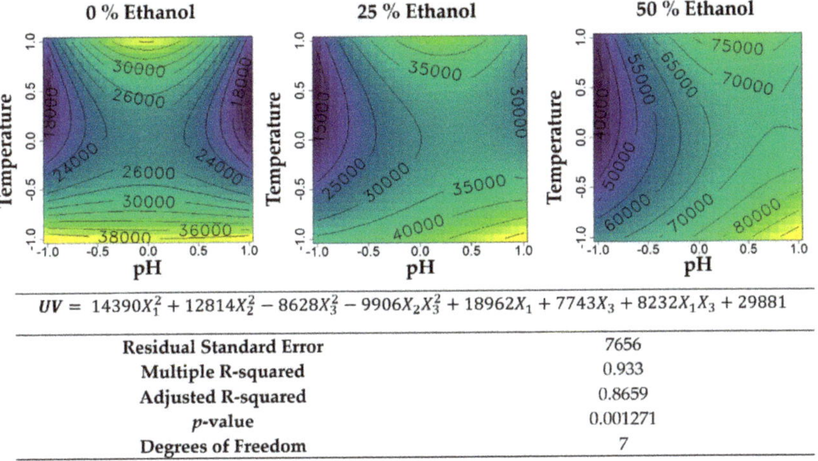

$$UV = 14390X_1^2 + 12814X_2^2 - 8628X_3^2 - 9906X_2X_3^2 + 18962X_1 + 7743X_3 + 8232X_1X_3 + 29881$$

Residual Standard Error	7656
Multiple R-squared	0.933
Adjusted R-squared	0.8659
***p*-value**	0.001271
Degrees of Freedom	7

Figure 7. Contour plots of the modelled function regarding the UV absorbance (per gram of extract) obtained during the second optimization of *Grateloupia turuturu* hydroethanolic SLE using a Box–Behnken design, in the presence of different concentrations of solvent (ethanol). For each plot, XX axis indicates the pH and YY axis indicates temperature. Color gradients represents the increase (yellow color) and decrease (blue color) in UV absorbance according to the tested variables. Below, the equation and goodness-of-fit statistical parameters for the variable UV absorbance (mL·cm^{-1}·g^{-1} extract) in the second optimization process, according to the ethanol concentration (X_1), temperature (X_2), and pH (X_3) are presented.

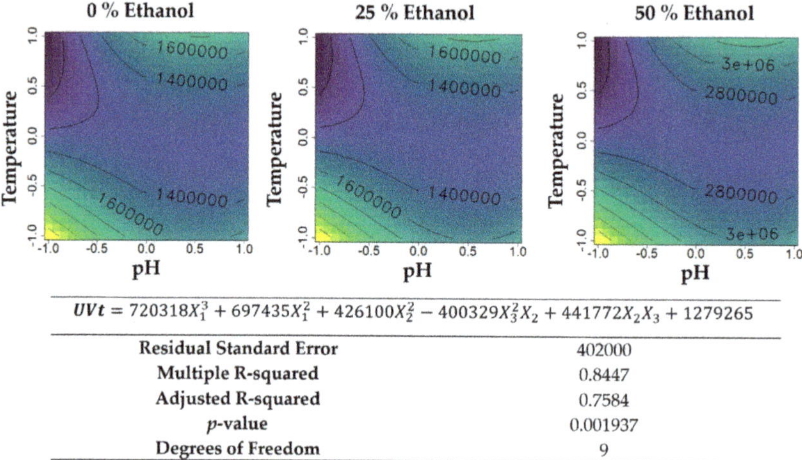

$$UVt = 720318X_1^3 + 697435X_1^2 + 426100X_2^2 - 400329X_3^2X_2 + 441772X_2X_3 + 1279265$$

Residual Standard Error	402000
Multiple R-squared	0.8447
Adjusted R-squared	0.7584
p-value	0.001937
Degrees of Freedom	9

Figure 8. Contour plots of the modelled function regarding the total UV absorbance (per gram of biomass) obtained during the second optimization of *Grateloupia turuturu* hydroethanolic SLE using a Box–Behnken design, in the presence of different concentrations of solvent (ethanol). For each plot, XX axis indicates the pH and YY axis indicates temperature. Color gradients represents the increase (yellow color) and decrease (blue color) in total UV absorbance according to the tested variables. Below, the equation and goodness-of-fit statistical parameters for the variable total UV absorbance (mL·cm^{-1}·g^{-1} biomass) in the second optimization process, according to the ethanol concentration (X_1), temperature (X_2), and pH (X_3) are presented.

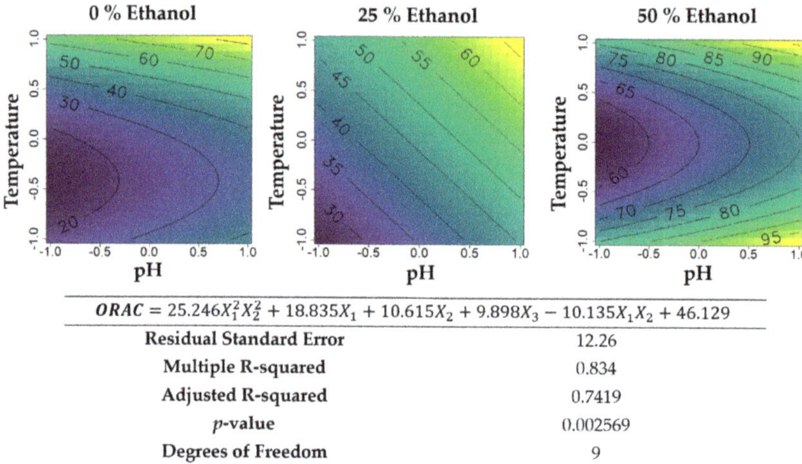

$$ORAC = 25.246X_1^2X_2^2 + 18.835X_1 + 10.615X_2 + 9.898X_3 - 10.135X_1X_2 + 46.129$$

Residual Standard Error	12.26
Multiple R-squared	0.834
Adjusted R-squared	0.7419
p-value	0.002569
Degrees of Freedom	9

Figure 9. Contour plots of the modelled function regarding the ORAC activity (per gram of extract) obtained during the second optimization of *Grateloupia turuturu* hydroethanolic SLE using a Box–Behnken design, in the presence of different concentrations of solvent (ethanol). For each plot, XX axis indicates the pH and YY axis indicates temperature. Color gradients represents the increase (yellow color) and decrease (blue color) in ORAC activity according to the tested variables. Below, the equation and goodness-of-fit statistical parameters for the variable antioxidant activity by ORAC method (μmol TE·g^{-1} extract) in the second optimization process, according to the ethanol concentration (X_1), temperature (X_2), and pH (X_3) are presented.

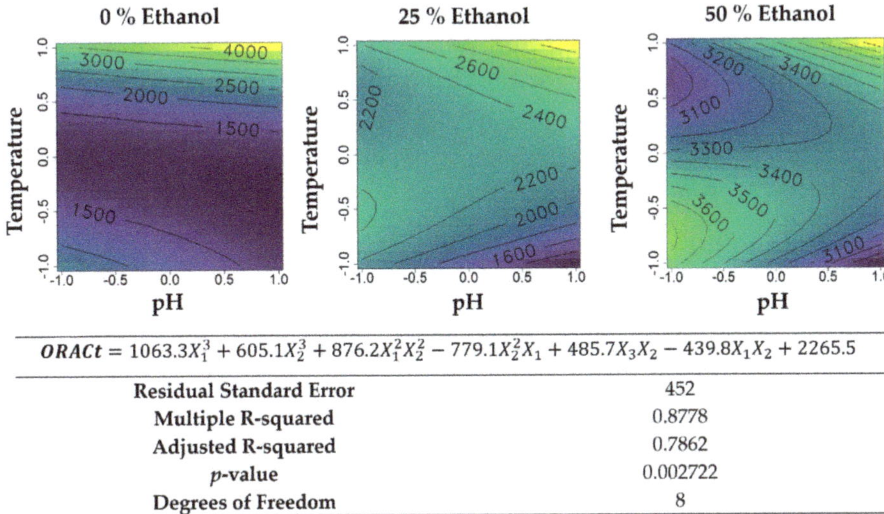

$$ORACt = 1063.3X_1^3 + 605.1X_2^3 + 876.2X_1^2X_2^2 - 779.1X_2^2X_1 + 485.7X_3X_2 - 439.8X_1X_2 + 2265.5$$

Residual Standard Error	452
Multiple R-squared	0.8778
Adjusted R-squared	0.7862
p-value	0.002722
Degrees of Freedom	8

Figure 10. Contour plots of the modelled function regarding the total ORAC activity (per gram of biomass) obtained during the second optimization of *Grateloupia turuturu* hydroethanolic SLE using a Box–Behnken design, in the presence of different concentrations of solvent (ethanol). For each plot, XX axis indicates the pH and YY axis indicates temperature. Color gradients represents the increase (yellow color) and decrease (blue color) in total ORAC activity according to the tested variables. Below, the equation and goodness-of-fit statistical parameters for the variable total ORAC activity (μmol TE·g^{-1} biomass) in the second optimization process, according to the ethanol concentration (X_1), temperature (X_2), and pH (X_3) are presented.

The total UV absorbance per gram of biomass (Figure 8) also increased with increasing percentages of ethanol, reaching the maximum value of 3,000,000 mL·cm^{-1}·g^{-1} biomass for the extractions with 50% ethanol (*v/v*). The aqueous extractions and the hydroethanolic extractions showed a similar profile, reaching the maximum values with acidic pH and lower temperatures.

Regarding the antioxidant activity per gram of extract (Figure 9), a slight increase of function values were found for the extractions, with 50% ethanol (*v/v*) (maximum of 95 μmol TE·g^{-1} extract and minimum of 60 μmol TE·g^{-1} extract) when compared with water-based extractions (maximum of 70 μmol TE·g^{-1} extract and minimum of 20 μmol TE·g^{-1} extract). For aqueous extractions, the conditions to obtain the maximum values are basic pH and higher temperatures (between 60 °C and 95°C) and for 50% ethanol (*v/v*) basic pH and temperatures at either of the extremes. For the hydroethanolic extractions with 25% ethanol (*v/v*), the higher the temperature and the pH, the higher antioxidant activity could be recovered.

The total ORAC activity per gram of biomass (Figure 10) showed to benefit from aqueous extractions, reaching the highest values between the tested conditions, using basic pH and high temperatures during the extraction procedure. For the hydroethanolic extractions, the highest values were obtained with the same conditions as for the aqueous extraction, but a skewed profile developed with a second relative maximum with acidic extractions at 25 °C.

4. Discussion

Global seaweed utilization is growing worldwide in response to the increasing demand for natural and functional ingredients [19]. Optimizations of seaweed extractions and high-throughput screening assays may contribute to find alternatives to the chemical industry, facing several challenges and limitations. In this study, crude extracts from *G. turuturu* produced by hydroethanolic SLE were optimized regarding extraction time, LSR, concentration of ethanol in water, solvent pH, and extraction temperature, using two RSM analysis with Box–Behnken designs [20].

Regarding the variable temperature, the requirements and cost for energy and equipment are lower at room temperature when compared with extraction at high temperatures. Working with neutral or natural solvent pH also excludes the need for additional resources to modify this variable, or for the maintenance of pH-related equipment deterioration. Likewise, lower LSR and extraction time are privileged conditions concerning cost effectiveness and environmental impact. Regarding ethanol percentage, due to the ease to evaporate and reuse it when compared to water, a balance between the increased cost of the solvent and the decreased cost of operation needs to be addressed. However, the availability and low toxicity of this solvent makes it readily accepted by the industry. Ethanol is also suitable for the use in cosmetics; it is sometimes even used as an ingredient. A balance between yield and specific activity of the extract (activity per mass unit) must be accounted when optimizing an extraction, since high yields and high specific activity are often inversely related and a maximum recovery of functional bioactivities can be found at intermediate values. Hence, biological activities coupled with good yields can supply good amounts of a final product. Nonetheless, lower yields with much higher activity, despite seemingly detrimental for the reduced quantity of extracts, may provide mixtures of compounds with less chemical complexity, turning the purification processes used for fine chemistry industries (such as cosmeceutical and pharmaceutical) cheaper or allow their incorporation in formulas at low rates, achieving the intended activity without compromising product safety. For these reasons, a comprehensive modelling of both yield and activities with extraction parameters is a dataset of paramount importance for the technology transfer of bioactive extracts from academy to industry.

The obtained extracts were used to evaluate antioxidant and UV-absorbing capacities, which are important features in the cosmeceutical industry. *Grateloupia*-specific SLE assumptions were made based on the literature: (1) dry biomass is mostly composed, in terms of massic contribution, of polysaccharides and proteins, which are almost exclusively soluble in water [21–24]; (2) polysaccharides are mostly agaran-carragenan hybrids (sulphated galactans) and thus have solubility/extractability highly conditioned by extraction variables such as time, LSR, temperature, and pH [22,23,25–30]; (3) mycosporine-like amino acids can be extracted with all the solvents used (water, ethanol 25%, and 50% (*v/v*)), possibly varying the relative content in each MAA with the solvent [27,31,32]; (4) proteins and MAAs are the main compounds responsible for UV-absorbance [4,27]; (5) ethanol inclusion in the solvent allows the recovery of compounds from highly polar (e.g., phenolic compounds) to medium polarity (carotenoids, sterols, fatty alcohols, etc.), all of which might contribute to antioxidant activity [33,34]; and (6) polysaccharides exhibit antioxidant activity, despite being below ethanol-extracted polar metabolites [22,35].

Three "slices" of cubic data at ethanol percentages 0, 25, and 50 were chosen to be represented as 2D contour plots (Figures 1–10). Ethanol was chosen to be the variable seen in slices because the effects found were not expected to lead to smooth, gradual changes in the parameters, as ethanol inclusion in the solvent leads to compositional changes that are more drastic than the variation of physicochemical parameters. Thus, despite being successfully modelled, ethanol effect will be discussed mostly as the difference between 0, 25, and 50% (*v/v*), discretely.

The first optimization addressed the effect of LSR, time, and ethanol percentage on the extraction of antioxidant and UV-absorbing molecules (Figures 1–5). Typically, LSR affects an extraction quadratically, having an optimal value, which decreases bilaterally [8,36]. Concerning time, it is usual to observe a steep increase in yield and other parameters of the extraction initially, approaching a plateau as time increases further, or a peak followed by decrease, in the case of sensitive bioactivities or low-solubility compounds abundance [36]. Ethanol inclusion, in the case of seaweeds, in which biomass extractives are mostly composed of polysaccharides and proteins, is expected to decrease yield and impact bioactivities differentially. In the case of *G. turuturu*, in this work, most of these trends were confirmed, with some exceptions.

Yield of *G. turuturu* hydroethanolic extractions (Figure 1) reached its maximum at 0% ethanol (water extraction), which was expected since water is capable of extracting the sulphated galactans of *G. turuturu*, and ethanol decreases solubility [25,26]. Moreover, water extracted the protein fraction from the biomass, which also had a significant massic contribution in its composition, where compounds (i.e., phycoerythrin) were present [4]. The effect pattern for LSR and time was maintained through different ethanol values; for the most part (the central range of values of time and LSR), yield increased linearly with LSR and time had almost no effect on extraction. Naturally, higher LSRs tended to favor the extraction of soluble compounds due to increased solvent availability. However, there were different LSR effects and time outside the mid-ranges. In extreme LSRs, time had a quadratic effect on extraction, with inverse trends depending on LSR. In low-range LSR ("shortage" of solvent), yield of short extraction times tended to decrease with an increase in time up to the 60 min of extraction and then increased again. It seems to be a counterintuitive phenomena, but it was repeatedly observed in previous works [37–39]. It may be that the initial contact of solvent with biomass promotd a rapid solubilization of certain extracellular biomass compounds and that the gradual extraction of intracellular compounds with time somehow promoted the precipitation of the formerly soluble, initally extracted ones. Then, with even greater extraction times, either by the continuation of intracellular compound extraction, or by the re-solubilization of the precipitates, yield increased again. In high-range LSR ("abundance" of solvent), the phenomenon inverted the monotony, maintaining the quadratic shape. Interestingly, for *G. turuturu*, this phenomena was of such proeminence that for 50% ethanol extractions, for extreme times of extraction, the trend of linear increase of yield with LSR inverts, and lower solvent volumes become more efficient.

The effect of LSR and time on UV absorbance of hydroethanolic extracts of *G. turuturu* is plotted in Figure 2. UV-absorbing compounds in this macroalgae species likely range from hydrophilic proteins and mycosporine-like aminoacids to polar organic compounds of phenolic or lipidic nature [34,35]. Concerning ethanol content in the solvent, the highest concentration (50% *v/v*) resulted in the most absorbing extracts, likely because this solvent composition can retrieve both types of highly-absorbing compounds (the polyphenols and the MAAs) with almost no extraction of carbohydrates (non-active, "activity-diluting" extractives). By reducing the amount of ethanol, UV absorbance decreased to less than half the massic extinction coefficient at 25% (*v/v*) ethanol compared to 50% (*v/v*) ethanol, at optimal times and LSR conditions. It might be because, at this value of ethanol, solubility of polyphenols and/or of specific MAAs decreased, and/or carbohydrate solubility increased. When approaching 0% (*v/v*) ethanol, an increase of UV absorbance was again observed, possibly because of increased protein content in the extracts. At a given ethanol percentage, LSR and time impacted the extraction of UV absorbing compounds differently. For instance, in water extracts, UV absorbance decreased with higher LSRs almost linearly, regardless of extraction time, and followed a quatradic curve (with a peak at intermediate extraction time) regardless of LSR. The decrease of activity with LSR might be correlated with the increase in yield; the extraction of more compounds in the yield increment cold be responsible for "diluting" the UV absorbing compounds with non-absorbing ones (likely carbohydrates). This suggests that the compounds responsible for UV absorption in the water extract are actually readily and highly water soluble, more than their co-extractives, which adds to the possibility of being proteins and/or MAAs. The occurrence of an optimal time of extraction, followed by a decrease of absorbance, suggests the degradation of these compounds with prolonged extractions. With the addition of ethanol, some skewing of the contours occurs at 25% (*v/v*) ethanol, where MAAs were likely the only absorbing compounds quantitatively extracted. Lower solvent volumes and higher extraction times led to minimum recovery, while higher solvent volumes with short extraction times maximized it. At 50% ehtanol extraction, more than 40 mL·g^{-1} of solvent and/or more than 100 min of extraction were required to achieve optimal recovery of UV absorbing compounds from *G. turuturu* biomass. By multiplying the specific UV absorbance of the extracts by their yield, one can further characterize the extraction in terms of total UV-absorbing capacity (Figure 3), and therefore final profitability should perform compound purification, as well high rates of inclusion of crude extracts.

In the case of hydroethanolic extraction, the total UV-absorbing capacity recovered became almost unimpacted by the ethanol percentage and the pattern of LSR effects and time became similar for different ethanol percentages. Nonetheless, either 0 or 50% (*v/v*) ethanol percentages were shown to best extrat solvents. Maximum recovery in thoses cases was attained using either low solvent volume for a long amount of time or high solvent volume for a mid-to-low amount of time. The similar patterns of the contour plots reinforced the idea that observations on the specific activity plots (Figure 7) were caused indirectly, in which the activity was "diluted" by non-active co-extractives, rather than directly, through differences in the amount of extracted UV-absorbing molecules. Thus, ethanol, LSR, and time might be good parameters to adjust in order to target higher-purity/potency extractions, especially for applications where purification is not meant to be performed.

The specific ORAC extract activity (Figure 4) was significantly impacted by all three variables under study. As expected, the more ethanol in the solvent, the higher the antioxidant activity. Alcohols were the preferred solvent for recovering antioxidant activity, as polar metabolites soluble in alcohols are often responsible for crude extract antioxidant activity [40]. In *G. turuturu*, antioxidant polyphenols, chlorophylls, and other hydroxyl-containing molecules occurred, such as polar carotenoids and tocopherols, all of which became more readily solubilized by the presence of ethanol in the solvent [19,33,41]. However, ORAC activity in water extracts from *G. turuturu* is worth studying, as sulphated carrageenans can present this activity [22,23] and their presence in a higher quantity might result in worthy application studies of these molecules. In the case of water extracts, both high volumes of solvent and long extraction times benefitted antioxidant recovery from *G. turuturu*. Actually, the 40 mL·g^{-1} and 100 min used in this study as upper limits of these variables did not decrease and rather increased. However, 10 mL·g^{-1} of solvent extracted for 100 min led to maximum ORAC activity, even higher than using 40 mL·g^{-1} for 100 min. If enough extraction time was employed, then increasing volume worsened the extract's specific antioxidant activity, likely because it allowed the extraction of non-active components that "diluted" the active ones. For 50% (*v/v*) ethanol, the same conditions (low volume and high time) resulted in maximum ORAC activity as well. However, for this solvent, higher volumes invariably resulted in same-or-poorer results, even at low times of extraction, which indicates that the recovery of the ethanol-soluble antioxidants was limited by the rate of mass transfer, and not by solubility.

Total ORAC activity (Figure 5) recovered per gram of biomass followed similar trends, in the case of 25% and 50% (*v/v*) ethanol. For water extracts, however, extraction under sub-optimal conditions regarding specific-activity, namely higher solvent volumes, regardless of time, because of an increase in yield, maximized antioxidant recovery to the point of obtaining maximum ORAC activity using 40 mL·g^{-1} for 20 min. If the extraction was made under optimal specific-ORAC activity, i.e., 10 mL·g^{-1} for 100 min, only 66% of total capacity available was recovered, despite being purer (higher per gram of extract). In any case, specific- and total-ORAC activities of *G. turuturu* hydroethanolic extracts had maximums in the limits of time and LSR tested, meaning that higher values of ORAC might be available in *G. turuturu* biomass if further optimization of these variables is performed.

The overall analysis of LSR, time, and ethanol percentage effects on the three parameters (yield, ORAC, and UV-absorption) of *G. turuturu* extracts suggested that ethanol percentage resulted in extracts fundamentally different at the chemical composition level, to a point where it should still be a variable under optimization in the study of temperature and pH effects. Thus, this variable was maintained in the second Box–Behnken design. For this second set of extractions, a LSR and time value was chosen by considering the overall effects observed in the first Box–Behnken, attempting to maximize yield, ORAC, and UV-absorption: 60 min of extraction and 40 mL·g^{-1}. Under these conditions, temperature and pH effects on different hydroethanolic extractions were characterized (Figures 6–10). Typically, for SLEs, temperature increases were expected to result when yield increased and more so when using water as a solvent. Solubility of organic substances tends to increase with temperature, and mass transfer phenomena become facilitated by kinetic energy possessed at higher temperatures [42,43]. The adjustment of pH is not something for which a trend is typically assigned, as it highly depends on the solvent and matrix being extracted; if pH changes results in higher or lower solubilities depends entirely on each molecule's characteristics [7]. For seaweed, however, because the matrix has a high content of polysaccharides and proteins, both macromolecules for which pH drastically alters their solubilities, pH adjustment is expected to result in very different extracts.

The yield of the extracts produced for the second optimization (Figure 6) revealed some unexpected effects. In the water extract, heating led to a decrease in yield that only reverted to an increase in the middle of the range (approximatley 60 °C), reaching a maximum yield at 95 °C, but had comparable values at room temperature. In the 25% (*v/v*) ethanol, the yield was not affected by temperature at a neutral pH. In the 50% (*v/v*) ethanol, the yield did increase with temperature, but in the middle of the range, it decreased to values comparable to room temperature. None of these three phenomena are in accordance with the SLE theory, but have been observed for seaweed extracts before; conditions of higher solvent availability/solubility sometimes result in lower yields [7]. In this case, it might be that, for water extracts, the compounds extracted at room temperature, plus the extra compounds extracted at 60 °C, end up interacting (for example, protein-polysaccharides adsorption), precipitating, and being removed along with the extracted biomass particles. The same phenomena might occur in the 50% (*v/v*) ethanol for the high-end temperatures tested. By modifying the pH, it is less odd that events like these occur or cease to occur. Side chains of proteins and polysaccharides become either totally protonated (in acid pH) or totally deprotonated (in alkaline pH) to a point where adsorption phenomena and solvation phenomena change significantly [42,43]. In *G. turuturu*, pH adjustment did not significantly impact the water-based extraction at room temperature, because whatever was being extracted had low molecular weight carbohydrates for which protonation was less relevant in terms of solubility than higher-molecular weight ones. At a near-boiling temperature, however, a linear increase of yield with pH suggested that alkalinity promoted extraction, which was expected since alkaline hydrolysis was one of the methods to dissolve cellulosic biomass and solvate their carbohydrates [44]. With ethanol in the solvent, this trend was reversed and we observed an almost-linear increase of yield and decreased pH. These events may be related to an increased ease of mass transfer due to a certain degree of acid and alkaline-hydrolysis of the cellulose cell walls. If that is the case, then in ethanol-containing solvents it has been demonstrated that acidic pH increased the solubility of *G. turuturu* metabolites, perhaps by disrupting the hydrogen bonds between proteins and such other solubles.

UV-absorbance of the extracts in the second Box–Behnken (Figure 7) followed a similar trend across the different ethanol percentages, which makes sense considering that in the initial optimization it was in the vicinity of 40 mg·mL^{-1} with 60 min of extraction, and the ethanol percentage had a low effect (pattern was similar). This reinforced that ethanol percentage, using this LSR and time, did not change extract composition in terms of UV-absorbing molecules (the proteins, the MAAs, and some polyphenols). However, like the first RSM, higher ethanol did promote higher absolute UV-absorbance values, possibly by decreasing the co-extraction of non-absorbing molecules. Both temperature and pH had significant effects on UV absorbance. Temperature followed a quadratic curve with a minimum absorbance at the middle of the range and, at the end, experience maximum near-RT and near-boiling. The compounds that were hypothesized to precipitate with the increased extraction of other molecules (observed in the yield) are possibly responsible for the UV absorbance (likely, the proteins). In the 50% (*v/v*) ethanol, even though the yield increased at mid temperature and decreased at higher ones, UV-absorbance was minimal at mid temperature increased with higher temperature. Chemical characterization of the extracts using chromatographic and spectroscopic techniques could assist in describing these mass tranfer nuances. The effect of pH was seemingly impacted by temperature, with neutral pH being better for UV absorbance recovery with hot extraction, following a quadratic curve, while room temperature extracts had no- to linear-effect with increased ethanol. The solubility of UV-absorbing proteins and MAAs might be impacted by pH since these molecules contain a certain balance of carboxylic and amine groups, which varies their state of protonation with pH [45]. Considering that yield is lower under alkaline conditions, it is likely that an increase in UV-absorbance of alkaline extracts at room temperature is, however, caused by a lower co-extraction of non-absorbing molecules, rather than a direct effect on the extraction of UV-absorving molecules.

Considering extraction yield and UV-absorbance together (total UV-absorbance, see Figure 8), it is clear that room temperature and acidic solvents, especially those containing ethanol, are the best extraction method (concerning pH and temperature) for UV-absorbing molecules recovery from *G. turuturu*. It is important to observe how significant the impact of pH adjustment is. In this study, when working at room temperature, the addition of HCl to the solvent allowed increments in UV-absorbing molecules a total recovery of 33% to 62% for 50% ethanol and 0% (*v/v*) ethanol, respectively. This represents a significant increase in profitability and therefore in the sustainability of industrially exploring this species of seaweed with a simple modification of the extraction protocol.

Finally, the antioxidant activity of the second Box–Behnken was addressed (Figure 9). Ethanol inclusion in the solvent of extraction, as expected, let to an increase in ORAC at room temperature. However, upon manipulation of temperature, pH, or both, ethanol led to a decrease in ORAC at low inclusion levels that only resumed the increasing trend with inclusions near 50% (*v/v*). It might be that pH and temperature manipulation leads to higher recovery of polysaccharides of different chemical natures, and that this recovery is different with high-, low-, and absent-ethanol. Regardless, at any given ethanol inclusion level, the effects of temperature and pH followed similar trends: an increase in pH led to a linear increase of ORAC; temperature, yield, and UV-absorbance had a minimum at the middle of the range, meaning either cold or hot extractions were privileged to medium-heat ones, possibly due to the phenomena of precipitation. The total ORAC analysis (Figure 10) revealed skewed patterns of effects and interactions between temperature and pH. However, for all ethanol percentages, ORAC-reactive antioxidants were extracted most at near-boiling temperatures, with little to no effect of pH, suggesting that higher temperature (under pressure) hydroethanolic extraction of *G. turuturu* may lead to even higher recoveries than those obtained in this study. Probably of a different chemical nature, with 50% (*v/v*) ethanol, cooler acidic extractions also resulted in the accumulation of substantial amounts of ORAC-active compounds. If these extracts are intended to be subjected to purification, despite the relatively low specific ORAC values at these conditions (pH between 4 and 7, temperature between 25 and 60 °C), then this less-energy consuming extraction mode might be of interest. On the other hand, for purification purposes, the highly active extracts of alkaline ethanol 50% (*v/v*) at room temperature fall short of high quantitative recovery. These extracts may be more suitable for direct applications of the crude extract given their inherent higher potency/concentration of active components.

5. Conclusions

The hydroethanolic extraction of *Grateloupia turuturu*, an invasive red macroalgae in the Iberian Peninsula, recognized for its potential for the cosmeceutical industry, has been thoroughly characterized regarding the effects of ethanol percentage, liquid-solid-ratio, time of extraction, temperature, and pH of the solvent in the yield of extract, UV-shielding potential, and antioxidant activity. Several critical observations regarding the equilibrium between high-value extracts and costs of operation were possible to perform. For instance, it was possible to demonstrate that for certain conditions, shorter extraction times, the extraction at room temperature, and the use of lower volumes of solvent are likely to be industrially more profitable than their counterparts, despite being a priori considered conditions of lower recovery of compounds to the extracts. More importantly, the data relating these five variables to these three effects is now available to the scientific community, which allows for further academic- and industry-oriented studies concerning biomass valorization.

Supplementary Materials: The following are available online at http://www.mdpi.com/2076-3417/10/15/5304/s1, Figure S1: Fit-real plots for the 10 models of response surface associated to the hydroethanolic SLE of *G. turuturu*, Figure S2: Statistical parameters associated to each factor/coefficient in the 10 models of response surface associated to the hydroethanolic SLE of *G. turuturu*, Table S1: Raw experimental data on yield, antioxidant activity and UV-shielding activity of the extracts produced in the two Box–Behnken designs performed of *G. turuturu* hydroethanolic SLE.

Author Contributions: Conceptualization, R.F. and M.F.F.L.; methodology, A.M.C.; software, R.F.; formal analysis, R.F.; investigation, R.F., A.M.C., and C.F.; writing—original draft preparation, R.F. and C.F.; writing—review and editing, A.M.C., S.C.N., and M.F.L.L.; supervision, S.C.N. and M.F.L.L.; project administration, M.F.F.L.; funding acquisition, M.F.L.L. All authors have read and agreed to the published version of the manuscript.

Funding: This study was supported by UID/MAR/04292/2019 with funding from FCT/MCTES through national funds and by a grant awarded to Rafael Félix (SFRH/ BD/139763/2018). The authors also wish to acknowledge the support of the European Union through EASME Blue Labs project AMALIA, Algae-to-MArket Lab IdeAs (EASME/EMFF/2016/1.2.1.4/03/SI2.750419), project VALORMAR (Mobilizing R&TD Programs, Portugal 2020) co-funded by COMPETE (POCI-01-0247-FEDER-024517), the Integrated Programme of SR&TD "Smart Valorization of Endogenous Marine Biological Resources Under a Changing Climate" (reference Centro-01-0145-FEDER-000018), co-funded by Centro 2020 program, Portugal 2020, European Union, and through the European Regional Development Fund, SAICTPAC/0019/2015—LISBOA-01-0145-FEDER-016405 Oncologia de Precisão: Terapias e Tecnologias Inovadoras (POINT4PAC).

Conflicts of Interest: The authors declare no conflict of interest. The funders had no role in the design of the study; in the collection, analyses, or interpretation of data; in the writing of the manuscript, or in the decision to publish the results.

References

1. Kim, S.K.; Ravichandran, Y.D.; Khan, S.B.; Kim, Y.T. Prospective of the cosmeceuticals derived from marine organisms. *Biotechnol. Bioprocess Eng.* **2008**, *13*, 511–523. [CrossRef]
2. Stengel, D.B.; Connan, S. Marine Algae: A Source of Biomass for Biotechnological Applications of the chapter. In *Natural Products From Marine Algae*; Humana: Louisville, KY, USA, 2015; Volume 1308.
3. Freitas, C.; Araújo, R.; Bertocci, I. Patterns of benthic assemblages invaded and non-invaded by Grateloupia turuturu across rocky intertidal habitats. *J. Sea Res.* **2016**, *115*, 26–32. [CrossRef]
4. Denis, C.; Massé, A.; Fleurence, J.; Jaouen, P. Concentration and pre-purification with ultrafiltration of a R-phycoerythrin solution extracted from macro-algae Grateloupia turuturu: Process definition and up-scaling. *Sep. Purif. Technol.* **2009**, *69*, 37–42. [CrossRef]
5. Araújo, R.; Violante, J.; Pereira, R.; Abreu, H.; Arenas, F.; Sousa-Pinto, I. Distribution and population dynamics of the introduced seaweed Grateloupia turuturu (Halymeniaceae, Rhodophyta) along the Portuguese coast. *Phycologia* **2011**, *50*, 392–402. [CrossRef]
6. Stiger-Pouvreau, V.; Zubia, M. Macroalgal diversity for sustainable biotechnological development in French tropical overseas territories. *Bot. Mar.* **2020**, *63*, 17–41. [CrossRef]
7. Santos, S.A.O.; Félix, R.; Pais, A.C.S.; Rocha, S.M.; Silvestre, A.J.D. The Quest for Phenolic Compounds from Macroalgae: A Review of Extraction and Identification Methodologies. *Biomolecules* **2019**, *9*, 847. [CrossRef]
8. Abidin, Z.Z.; Biak, D.R.A.; Yusoff, H.M.; Harun, M.Y. Solid-Liquid Extraction in Biorefinery. In *Separation and Purification Technologies in Biorefineries*; Ramaswamy, S., Huang, H.-J., Ramarao, B.V., Eds.; Wiley: Hoboken, NJ, USA, 2013; pp. 351–374.
9. Do, Q.D.; Angkawijaya, A.E.; Tran-Nguyen, P.L.; Huynh, L.H.; Soetaredjo, F.E.; Ismadji, S.; Ju, Y.-H. Effect of extraction solvent on total phenol content, total flavonoid content, and antioxidant activity of Limnophila aromatica. *J. Food Drug Anal.* **2014**, *22*, 296–302. [CrossRef]
10. Cardoso, I.; Cotas, J.; Rodrigues, A.; Ferreira, D.; Osório, N.; Pereira, L. Extraction and Analysis of Compounds with Antibacterial Potential from the Red Alga Grateloupia turuturu. *J. Mar. Sci. Eng.* **2019**, *7*, 220. [CrossRef]
11. Sady, S.; Matuszak, L.; Błaszczyk, A. Optimisation of ultrasonic-assisted extraction of bioactive compounds from chokeberry pomace using response surface methodology. *Acta Sci. Pol. Technol. Aliment.* **2019**, *18*, 249–256. [CrossRef]
12. Marcos, A.B.; Ricardo, E.S.; Eliane, P.O.; Leonardo, S.V.; Luciane, A.E. Response surface methodology (RSM) as a tool for optimization in analytical chemistry. *Talanta* **2008**, *76*, 965–977. [CrossRef]
13. Maciel, O.M.C.; Tavares, R.S.N.; Caluz, D.R.E.; Gaspar, L.R.; Debonsi, H.M. Photoprotective potential of metabolites isolated from algae-associated fungi Annulohypoxylon stygium. *J. Photochem. Photobiol. B Biol.* **2018**, *178*, 316–322. [CrossRef] [PubMed]
14. Dávalos, A.; Gómez-Cordovés, C.; Bartolomé, B. Extending Applicability of the Oxygen Radical Absorbance Capacity (ORAC-Fluorescein) Assay. *J. Agric. Food Chem.* **2004**, *52*, 48–54. [CrossRef] [PubMed]
15. R Core Team R: A Language and Environment for Statistical Computing 2013. Available online: https://www.R-project.org/ (accessed on 24 April 2020).

16. Lenth, R.V. Response-Surface Methods in {R.}, Using {rsm}. *J. Stat. Softw.* **2009**, *32*, 1–17. [CrossRef]
17. Garnier, S. viridis: Default Color Maps from "matplotlib" 2018. Available online: https://cran.r-project.org/web/packages/viridis/index.html/ (accessed on 26 April 2020).
18. RStudio Team RStudio: Integrated Development for R. 2015. Available online: https://https://rstudio.com// (accessed on 2 April 2020).
19. Tiwary, B.K. Seaweed Sustainability—Food and Non-Food Applications. In *Seaweed Sustainability*; Tiwari, B.K., Troy, D.J., Eds.; Academic Press: Cambridge, MA, USA, 2015.
20. Cavazzuti, M. *Optimization Methods: From Theory to Design Scientific and Technological Aspects in Mechanics*; Springer: Cham, Switzerland, 2013.
21. Yang, Z.; Yang, H.H.; Yang, H.H. Characterisation of rheology and microstructures of κ-carrageenan in ethanol-water mixtures. *Food Res. Int.* **2018**, *107*, 738–746. [CrossRef]
22. Ye, D.; Jiang, Z.; Zheng, F.; Wang, H.; Zhang, Y.; Gao, F.; Chen, P.; Chen, Y.; Shi, G. Optimized Extraction of Polysaccharides from Grateloupia livida (Harv.) Yamada and Biological Activities. *Molecules* **2015**, *20*, 16817–16832. [CrossRef]
23. Tang, L.; Chen, Y.; Jiang, Z.; Zhong, S.; Chen, W.; Zheng, F.; Shi, G. Purification, partial characterization and bioactivity of sulfated polysaccharides from Grateloupia livida. *Int. J. Biol. Macromol.* **2017**, *94*, 642–652. [CrossRef]
24. Zuhong, X.; Shengyao, S.; Zhien, L.; Yucai, G.; Xhingjun, Z. Study on carrageenan from Grateloupia filicina, Grateloupia filicina var. lomentaria and Pachymeniopsis elliptica. *Oceanol. Limnol. Sin.* **1996**, *27*, 499–504.
25. Denis, C.; Morançais, M.; Li, M.; Deniaud, E.; Gaudin, P.; Wielgosz-Collin, G.; Barnathan, G.; Jaouen, P.; Fleurence, J. Study of the chemical composition of edible red macroalgae Grateloupia turuturu from Brittany (France). *Food Chem.* **2010**, *119*, 913–917. [CrossRef]
26. Guo, M.Q.; Hu, X.; Wang, C.; Ai, L. Polysaccharides: Structure and Solubility. In *Solubility of Polysaccharides*; Intechopen Ltd.: London, UK, 2017.
27. Torres, P.; Santos, J.P.; Chow, F.; Pena Ferreira, M.J.; dos Santos, D.Y.A.C. Comparative analysis of in vitro antioxidant capacities of mycosporine-like amino acids (MAAs). *Algal Res.* **2018**, *34*, 57–67. [CrossRef]
28. Miller, I.J. The structure of polysaccharides from selected New Zealand species of Grateloupia. *Bot. Mar.* **2005**, *48*, 157–166. [CrossRef]
29. Sen, A.K.; Das, A.K.; Sarkar, K.K.; Siddhanta, A.K.; Takano, R.; Kamei, K.; Hara, S. An agaroid-carrageenan hybrid type backbone structure for the antithrombotic sulfated polysaccharide from Grateloupia indica Boergensen (Halymeniales, Rhodophyta). *Bot. Mar.* **2002**, *45*, 331–338. [CrossRef]
30. Wang, S.C.; Bligh, S.W.A.; Shi, S.S.; Wang, Z.T.; Hu, Z.B.; Crowder, J.; Branford-White, C.; Vella, C. Structural features and anti-HIV-1 activity of novel polysaccharides from red algae Grateloupia longifolia and Grateloupia filicina. *Int. J. Biol. Macromol.* **2007**, *41*, 369–375. [CrossRef] [PubMed]
31. Chaves-Peña, P.; de la Coba, F.; Figueroa, F.L.; Korbee, N. Quantitative and Qualitative HPLC Analysis of Mycosporine-Like Amino Acids Extracted in Distilled Water for Cosmetical Uses in Four Rhodophyta. *Mar. Drugs* **2019**, *18*, 27. [CrossRef] [PubMed]
32. Huovinen, P.; Gómez, I.; Figueroa, F.L.; Ulloa, N.; Morales, V.; Lovengreen, C. Ultraviolet-absorbing mycosporine-like amino acids in red macroalgae from Chile. *Bot. Mar.* **2004**, *47*, 21–29. [CrossRef]
33. Shalaby, E.A. Algae as promising organisms for environment and health. *Plant Signal. Behav.* **2011**, *6*, 1338–1350. [CrossRef] [PubMed]
34. Jiang, Z.; Chen, Y.; Yao, F.; Chen, W.; Zhong, S.; Zheng, F.; Shi, G. Antioxidant, Antibacterial and Antischistosomal Activities of Extracts from Grateloupia livida (Harv). Yamada. *PLoS ONE* **2013**, *8*, e80413. [CrossRef]
35. Athukorala, Y.; Lee, K.W.; Song, C.; Ahn, C.B.; Shin, T.S.; Cha, Y.J.; Shahidi, F.; Jeon, Y.J. Potential antioxidant activity of marine red alga grateloupia filicina extracts. *J. Food Lipids* **2003**, *10*, 251–265. [CrossRef]
36. Meireles, M.A.A. *Extracting Bioactive Compounds for Food Products: Theory and Applications*; CRC Press: Boca Raton, FL, USA, 2009.
37. Chemat, F.; Strube, J. *Green Extraction of Natural Products: Theory and Practice*; Wiley-VCH: Weinheim, Germany, 2014.
38. Li, H.; Zhang, H.; Zhang, Z.; Cui, L. Optimization of ultrasound-assisted enzymatic extraction and in vitro antioxidant activities of polysaccharides extracted from the leaves of Perilla frutescens. *Food Sci. Technol.* **2020**, *40*, 36–45. [CrossRef]

Appl. Sci. **2020**, *10*, 5304

39. Che Sulaiman, I.S.; Basri, M.; Fard Masoumi, H.R.; Chee, W.J.; Ashari, S.E.; Ismail, M. Effects of temperature, time, and solvent ratio on the extraction of phenolic compounds and the anti-radical activity of Clinacanthus nutans Lindau leaves by response surface methodology. *Chem. Cent. J.* **2017**, *11*, 54. [CrossRef]

40. Sultana, B.; Anwar, F.; Ashraf, M. Effect of extraction solvent/technique on the antioxidant activity of selected medicinal plant extracts. *Molecules* **2009**, *14*, 2167–2180. [CrossRef]

41. Garcia-Vaquero, M.; Rajauria, G.; O'Doherty, J.V.; Sweeney, T. Polysaccharides from macroalgae: Recent advances, innovative technologies and challenges in extraction and purification. *Food Res. Int.* **2017**, *99*, 1011–1020. [CrossRef]

42. Joana Gil-Chávez, G.; Villa, J.A.; Fernando Ayala-Zavala, J.; Basilio Heredia, J.; Sepulveda, D.; Yahia, E.M.; González-Aguilar, G.A. Technologies for Extraction and Production of Bioactive Compounds to be Used as Nutraceuticals and Food Ingredients: An Overview. *Compr. Rev. Food Sci. Food Saf.* **2013**, *12*, 5–23. [CrossRef]

43. Gertenbach, D.D. Solid-liquid extraction technologies for manufacturing nutraceuticals. In *Functional Foods: Biochemical and Processing Aspects*; CRC Press: Boca Raton, FL, USA, 2016.

44. Al-Nahdi, Z.M.; Al-Alawi, A.; Al-Marhobi, I. The Effect of Extraction Conditions on Chemical and Thermal Characteristics of Kappa-Carrageenan Extracted from Hypnea bryoides. *J. Mar. Biol.* **2019**, *2019*, 1–10. [CrossRef]

45. Carreto, J.I.; Carignan, M.O. Mycosporine-like amino acids: Relevant secondary metabolites. chemical and ecological aspects. *Mar. Drugs* **2011**, *9*, 387–446. [CrossRef]

Article

Cosmeceutical Potential of *Grateloupia turuturu*: Using Low-Cost Extraction Methodologies to Obtain Added-Value Extracts

Carina Félix [1,*], Rafael Félix [1,2], Ana M. Carmona [1], Adriana P. Januário [1], Pedro D.M. Dias [1], Tânia F.L. Vicente [1], Joana Silva [1], Celso Alves [1], Rui Pedrosa [1], Sara C. Novais [1] and Marco F.L. Lemos [1,*]

1 MARE–Marine and Environmental Sciences Centre, ESTM, Politécnico de Leiria, 2520-641 Peniche, Portugal; rafael.felix@ipleiria.pt (R.F.); 4160376@my.ipleiria.pt (A.M.C.); adriana.p.januario@ipleiria.pt (A.P.J.); pedro.d.dias@ipleiria.pt (P.D.M.D.); tania.vicente@ipleiria.pt (T.F.L.V.); joana.m.silva@ipleiria.pt (J.S.); celso.alves@ipleiria.pt (C.A.); rui.pedrosa@ipleiria.pt (R.P.); sara.novais@ipleiria.pt (S.C.N.)
2 REQUIMTE/LAQV, Laboratório de Farmacognosia, Faculdade de Farmácia, Universidade do Porto, 4050-313 Porto, Portugal
* Correspondence: carina.r.felix@ipleiria.pt (C.F.); marco.lemos@ipleiria.pt (M.F.L.L.)

Abstract: The invasive macroalga *Grateloupia turuturu* is known to contain a diversity of bioactive compounds with different potentialities. Among them are compounds with relevant bioactivities for cosmetics. Considering this, this study aimed to screen bioactivities with cosmeceutical potential, namely, antioxidant, UV absorbance, anti-enzymatic, antimicrobial, and anti-inflammatory activities, as well as photoprotection potential. Extractions with higher concentrations of ethanol resulted in extracts with higher antioxidant activities, while for the anti-enzymatic activity, high inhibition percentages were obtained for elastase and hyaluronidase with almost all extracts. Regarding the antimicrobial activity, all extracts showed to be active against *E. coli*, *S. aureus*, and *C. albicans*. Extracts produced with higher percentages of ethanol were more effective against *E. coli* and with lower percentages against the other two microorganisms. Several concentrations of each extract were found to be safe for fibroblasts, but no photoprotection capacity was observed. However, one of the aqueous extracts was responsible for reducing around 40% of the nitric oxide production on macrophages, showing its anti-inflammatory potential. This work highlights *G. turuturu*'s potential in the cosmeceutical field, contributing to the further development of natural formulations for skin protection.

Keywords: bioactive compounds; invasive seaweed; skincare; antioxidant activity; antimicrobial activity; cytotoxicity; anti-enzymatic activity; anti-inflammatory activity

Citation: Félix, C.; Félix, R.; Carmona, A.M.; Januário, A.P.; Dias, P.D.M.; Vicente, T.F.L.; Silva, J.; Alves, C.; Pedrosa, R.; Novais, S.C.; et al. Cosmeceutical Potential of *Grateloupia turuturu*: Using Low-Cost Extraction Methodologies to Obtain Added-Value Extracts. *Appl. Sci.* **2021**, *11*, 1650. https://doi.org/10.3390/app11041650

Academic Editor: Claudia Clelia Assunta Juliano

Received: 15 January 2021
Accepted: 8 February 2021
Published: 12 February 2021

Publisher's Note: MDPI stays neutral with regard to jurisdictional claims in published maps and institutional affiliations.

1. Introduction

Marine organisms' environments are known to be deeply demanding due to competition and extreme conditions, forcing them to develop defense mechanisms and produce secondary metabolites to survive and protect themselves against external threats [1,2]. These produced compounds make marine organisms great sources of bioactive compounds with a myriad of applications. Among them, macroalgae are one of the most ecologically and economically relevant marine resources to obtain this type of compound, having in their constitution fibers, proteins, amino acids, minerals, polyunsaturated fatty acids, and vitamins [2].

Grateloupia turuturu (Yamada, 1941) is the largest edible red macroalga in the world. It is native to Korea and Japan and was classified as an invasive species in the Atlantic Ocean, being the first report in Portugal from 1997 [3]. It is typically characterized by a high content of carbohydrates (such as sulfated polysaccharides, known antioxidants, and antimicrobials), proteins (such as chromoproteins, with known antioxidant activity), and secondary metabolites (such as mycosporine-like amino acids (MAAs), known for their

UV-shielding activity) and a low content of lipids [4–7]. The presence of such compounds is responsible for avoiding damages caused by the frequent exposure to UV radiation and high oxidative stress levels, typically found in mid-intertidal areas and intertidal pools, where this species is mostly located [8]. Developing extraction methodologies that are industrially feasible will create added-value extracts (such as cosmeceutical ingredients) that can turn to be an opportunity to promote the harvesting of this species, with positive consequences for the local invaded environments. For that, solid–liquid extraction (SLE) is one of the most suitable solutions due to the ease to up-scale the production. In fact, for the specific case of *G. turuturu*, solid–liquid extraction using ethanol and water as solvents is an already optimized method to extract the main bioactive compounds [9].

Personal care and image are receiving more attention every day, resulting in an unprecedented increase in cosmetic products use [1]. In 2016, Europeans spent a total of EUR 77 billion in this field, followed by the United Sates with EUR 64 billion and Brazil with EUR 24 billion [2]. The current concept of beauty includes healthy skin and a young appearance. Thus, the formulations to control the signs of aging are one of the industry's biggest demands [10].

Skin aging is a natural and progressive process that is influenced by two main factors: intrinsic factors, such as genetics and physiological alterations, and extrinsic factors, such as environment, exposure to UV radiation or even smoking [1,2,11]. The signs of skin aging include thinning, fragility and continuous losses of elasticity of the skin, as well as the inability to maintain hydration, resulting in the formation of wrinkles [2,10]. In this process, the antioxidant defense system loses the capacity to block reactive oxygen species (ROS), leading to oxidative stress [10]. Together with reactive nitrogen species (RNS), they participate in regular cellular functions, being responsible for several regulatory mechanisms of cells to protect them against oxidative stress [2]. However, an overproduction of these molecules can play a different role, inducing damages in different cell structures, such as membranes, DNA, proteins and lipids, among others [2]. Thus, products able to reduce the symptoms of aging and consequently increase the quality of life and the self-esteem of consumers are among the most wanted, being used on a daily basis by millions of people [11]. Currently, an increased demand for natural solutions by customers [1,10] that replace the use of synthetic chemicals exists, due to the latter having high costs and being more pollutant and less sustainable, while also being perceived by the public as less safe.

Therefore, the main goal of this study is the evaluation of the bioactivities of several extracts from *Grateloupia turuturu*, taking into consideration the solvents used in the extraction procedure, with potential to be applied in natural skincare formulations, adding value to this species. For that, antioxidant, UV absorbance, anti-enzymatic, antimicrobial, and anti-inflammatory bioactivities are evaluated, as well as the cytotoxicity of extracts in fibroblasts and their photoprotection potential.

2. Materials and Methods

2.1. Seaweed Collection

The red seaweed *Grateloupia turuturu* was collected at Aguda Beach in Arcozelo, Portugal (41.054826, −8.656865), in July of 2017. The collected biomass was sorted for epibionts and then dried in a wind tunnel at 25 °C. The dried biomass was milled to flour-like powder (particle size 150 ± 50 μm) and stored under vacuum in the dark, at room temperature, until use.

2.2. Seaweed Extracts

Optimal conditions for 4 hydroethanolic solid–liquid extracts of *G. turuturu* and 2 aqueous extracts were selected (see Table 1) according to the optimization of the extraction process performed by Félix and co-workers [9], and their extraction methodology was followed. Briefly, two optimization assays were performed using a response surface methodology with a Box–Benhken design. Firstly, the solid-liquid ratio (SLR), the time of extraction (min) and the ethanol percentage were addressed. Then, using the results obtained for these 3 independent parameters, the influence of the extraction temperature (°C), pH and ethanol percentage was evaluated [9].

Table 1. Selected extracts and respective extraction conditions: temperature, pH, percentage of ethanol, time and solid–liquid ratio.

Extracts	Temperature (°C)	pH	% EtOH	Time (min)	SLR
E1	30	9	50	60	1:40
E2	100	9	50	60	1:40
E3	100	7	25	100	1:10
E4	20	4	25	100	1:10
E5	20	9	0	20	1:40
E6	100	9	0	20	1:40

For the production of the selected extracts of *Grateloupia turuturu*, 5 g of biomass with the selected volume of solvent was mixed under constant magnetic stirring and thermostatized during the selected time of extraction. Each extract was then centrifuged for 5 min at $10,000\times$ g and the obtained supernatant was filtered using filter paper (Whatmann no. 1). The evaporation of the extracts was performed under reduced pressure at 40 °C and then desiccated at room temperature using a vacuum concentrator (Vacufuge, Eppendorf, Germany). Yield of dry extracts was calculated (g extract·g^{-1} biomass) and then they were resuspended: aqueous extracts were resuspended in water at 25 mg·mL^{-1}; 25% (*v/v*) ethanol extracts were resuspended in 25% (*v/v*) DMSO in water at 50 mg·mL^{-1}; and 50% (*v/v*) ethanol extracts were resuspended in 50% (*v/v*) DMSO in water at 100 mg·mL^{-1}.

2.3. Antioxidant Activity and UV Absorbance

The antioxidant activity was measured by ORAC assay, according to Félix and colleagues [9] and Dávalos and co-workers [12]. Briefly, a Trolox stock solution (VWR, Radnor, PA, USA) was used to prepare the dilutions from 8 to 0.5 µM. The obtained extracts were tested at 1 mg·mL^{-1} (diluted in 75 mM phosphate buffer). A fluorescein solution at 70 nM was used and the AAPH (2,2′-Azobis(2-methylpropionamidine) dihydrochloride) reagent (Sigma, Darmstadt, Germany) at 12 mM was prepared. A total of 20 µL of each sample was used and 120 µL of a fluorescein solution (70 nM) (Sigma, Darmstadt, Germany) was added to all samples in a 96-well black microplate (Greiner, Austria), including the standard curve. Phosphate buffer, at 75 mM, was used as control. Fluorescence was read for 15 min with a 1-min interval at an excitation wavelength of 485 nm and an emission wavelength of 525 nm in a microplate reader (Synergy H1, Biotek, Winooski, VT, USA) at 37 °C. After the incubation period, 60 µL of AAPH at 37 °C was added. The fluorescence was read for 80 min with 1-min intervals. Results were expressed as µmol of Trolox equivalents per gram of extract (µmol TE·$^{-1}$ ext) and are reported as the mean of three replicates and standard deviation.

UV absorption was also performed according to [9]. Briefly, 200 µL of each extract (0.1 mg·mL^{-1}) was added to a 96-well microplate for UV readings (Greiner UV-Star®, Kremsmünster, Austria) as well as the respective blanks. The absorbance was read between 280 and 400 nm (Synergy H1, Biotek, Winooski, VT, USA). The integral of the absorbance (Abs) was used to calculate the area under the curve (AUC), which was reported as the mean of three replicates and standard deviation.

2.4. Anti-Enzymatic Activity

2.4.1. Elastase Inhibition

The inhibition of elastase activity of the six extracts was performed using the EnzChek® Elastase Assay Kit (Invitrogen, Carlsbad, CA, USA) following the manufacturer's instructions. A total of 50 µL of each extract at 2 mg·mL^{-1} was incubated with 50 µL of DQ-elastin from bovine neck ligament, BODIPY FL conjugate, in reaction buffer. Enzymatic release of fluorescent signal from DQ-elastin by elastase was quantitated using a fluorescent microplate reader (Synergy H1, Biotek, Winooski, VT, USA) at 486 nm excitation and 525 nm emission. To stop the enzymatic activity, N-Methoxysuccinyl-Ala-Ala-Pro-Val-chloromethyl ketone was added to the reaction buffer at a final concentration of 0.25 mg·mL^{-1}. Elastase from pig pancreas was used at a final concentration of 0.025 mg·mL^{-1}. Results are expressed as percentage of elastase inhibition.

2.4.2. Hyaluronidase Inhibition

The inhibition of hyaluronidase activity of the 6 extracts was performed according to Madan et al. and Adamczyk and colleagues [13,14], with some modifications. Hyaluronidase solution (4 U·mL^{-1}) was prepared using a stock solution containing sodium phosphate buffer (200 mM, pH 7, 37 °C), 77 mM sodium chloride and 0.01% BSA. In a 1.5 mL tube, 200 μL of hyaluronidase solution and 25 μL of extract at 5 mg·mL^{-1} were incubated at 37 °C, for 10 min. Then, 100 μL hyaluronic acid solution (prepared in 300 mM of sodium phosphate monobasic solution at 0.06%) was added and the mixture was incubated at 37 °C, for 75 min. After the incubation period, 1 mL of acidic BSA (0.1% bovine serum albumin, 24 mM sodium acetate and 79 mM acetic acid, pH 3.75) was added and mixed by inversion, transferred to 96-well microplates and incubated for 15 min at room temperature. The absorbance was measured at 600 nm, in a microplate reader (Epoch2, Biotek, Winooski, VT, USA), and the data are presented as inhibition percentage.

2.5. Antimicrobial Activity

Antimicrobial activity of the six extracts of *G. turuturu* was evaluated through the microdilution technique [15,16] with slight modifications, using a fungal strain of *Candida albicans* (DSM-1386), the Gram-negative bacterium *Escherichia coli* (DSM-1103) and the Gram-positive bacterium *Staphylococcus aureus* (DSM-1104). The *C. albicans* two-day grown culture (Yeast and Mold Agar; VWR, cc) and the *E. coli* and *S. aureus* over-night grown cultures (Nutrient Agar; Sigma, Germany) were dissolved in saline solution (0.85% NaCl; Merck Millipore, Germany) and adjusted to a concentration of 1×10^7 (for bacteria) and 2×10^4 CFU·mL^{-1} (for fungus). The final inoculum concentrations on the microplates were 5×10^5 (bacteria) and 1×10^3 CFU.mL^{-1} (fungus), using Mueller-Hinton broth 2 (Sigma, Darmstadt, Germany) and RPMI-1640 (Sigma, Darmstadt, Germany). The positive control of inhibition used for *E. coli* was Ciprofloxacin (4 μg·mL^{-1}; Sigma, Darmstadt, Germany), for S. aureus was Tetracycline (16 μg·mL^{-1}, Sigma, Darmstadt, Germany) and Amphotericin B (4 μg·mL^{-1}; Sigma, Darmstadt, Germany) was used for the *C. albicans* positive control; 4% *(v/v)* DMSO (Dimethyl sulfoxide; Carlo Erba, Spain) was used as negative control of microbial inhibition. *Grateloupia turuturu* extracts were tested at 0.0075, 0.75, 1.5 and 3 mg·mL^{-1} (diluted in phosphate saline buffer), using sterile round-bottom microplates (Thermo Scientific, Waltham, MA, EUA). For *E. coli* and *S. aureus*, the incubation period was 20 h at 35 °C, and for *C. albicans*, it was 48 h at 35 °C. After this time, the optical density (DO) was measured at 625 (bacteria) or 530 nm (fungus), in a microplate reader (Epoch2, BioTek, Winooski, VT, USA). The test was performed using 3 independent assays. Results are expressed in percentage of bacterial growth inhibition.

2.6. Photoprotection Activity

A 3T3 cell line (DSMZ–ACC 173, mouse fibroblasts) was grown and maintained according to supplier's instructions. The cytotoxicity of the extracts was evaluated using the neutral red method described by Repetto et al. with slight modifications [17]. The 96-well microplates containing 5×10^4 cells/well were incubated at 37 °C in 5% CO_2 for 24 h in Dulbecco's modified Eagle medium (DMEM) (Sigma, Darmstadt, Germany), 10% FBS (Biowest, Nuaillé, France). Cells were treated for 24 h with extracts (1:1 in DMEM, 10% FBS). A dose–response evaluation with eight different concentrations of each extract was performed (0.01, 0.062, 0.125, 0.25, 0.5, 1, 2 and 4 mg·mL^{-1} in phosphate-buffered saline [PBS]) in order to find the non-cytotoxic concentrations for the cells. After the incubation period, the medium was removed by aspiration and washed with 100 μL of PBS. After that,

100 µL of DMEM with 5% FBS, without phenol red and supplemented with neutral red (40 µg·mL^{-1} in PBS) (Sigma, Darmstadt, Germany), was added to each well to assess cell viability. The microplates were incubated at 37 °C in 5% CO_2 for 4 h and then washed with PBS. After aspiration, 100 µL of desorption solution containing glacial acetic acid, ultrapure water and absolute ethanol (1:49:50) was added and the microplates were agitated until complete homogenization. The absorbance was read at 540 nm wavelength in a microplate spectrophotometer (Epoch2, BioTek, Winooski, VT, USA). PBS supplemented with the respective concentration of DMSO (vehicle) present in each sample and DMEM medium were used as controls. Data presented are the result of 3 independent replicas.

Knowing the non-cytotoxic concentrations of each extract, the concentration closest to 100% of cell viability was selected to perform a phototoxicity assay. The same 3T3 cell line was used and the assay was performed according to the OECD "Guidelines for Testing of Chemicals-In Vitro 3T3 NRU Phototoxicity Test" [18], with slight modifications. For the photoprotection evaluation of the extracts against UV radiation, the 96-well microplates containing 5×10^4 cells/well were incubated at 37 °C in 5% CO_2 for 24 h in DMEM, 10% FBS. After that period, the medium was removed and cells were treated for 1 h with 100 µL of each extract (E1—0.01 mg·mL^{-1}, E2—0.062 mg·mL^{-1}, E3—0.5 mg·mL^{-1}, E4—0.5 mg·mL^{-1}, E5—0.25 mg·mL^{-1}, and E6—0.5 mg·mL^{-1}, diluted in PBS) and then exposed for 40 min to UVA radiation (200 mJ/cm^2) using a sun simulator chamber (UVA Cube 400, SOL500, Hönle UV Technology, Gräfelfing, Germany) equipped with a UVA filter (H1) (Hönle UV Technology, Gräfelfing, Germany) and a UVA sensor (FS UV-A D0, Hönle UV Technology, Gräfelfing, Germany) with a spectral range of 330–400 nm. After the exposure period, extracts were removed and the wells were washed with PBS, substituted by new medium and incubated at 37 °C in 5% CO_2 for 24 h. After the incubation period, the medium was removed by aspiration and washed with 100 µL of PBS and the cytotoxicity was evaluated following the neutral red assay described above. PBS supplemented with the respective concentration of DMSO present in each sample and DMEM medium were used as positive controls and DMSO and empty wells were used as negative controls. Each condition was tested using 6 technical replicates and 3 independent assays.

2.7. NO Measurement

A RAW 264.7 cell line (ATCC-TIB 71, mouse macrophages) was grown and maintained according to the supplier's instructions.

The effect of different concentrations of the macroalgal extracts on cell toxicity was determined using the 3-(4,5-dimethylthiazol-2-yl)-2,5-diphenyl tetrazolium bromide (MTT) assay according to Bahiense and colleagues, with slight modifications [19]. RAW 264.7 cells were treated for 24 h with extracts at increasing concentrations (0.01, 0.062, 0.125, 0.25, 0.5, 1, 2 and 4 mg·mL^{-1}) in PBS. After the incubation period, the medium was removed by aspiration and washed with 100 µL of PBS. After that, 100 µL of DMEM with 5% FBS, without phenol red and supplemented with MTT solution (0.5 mg·mL^{-1} in PBS) (Sigma, Darmstadt, Germany), was added to each well to assess cell viability. The microplates were incubated at 37 °C in 5% CO_2 for 4 h and then washed with PBS. After aspiration, 100 µL of DMSO was added and the microtiter plates were agitated for a few minutes and kept in the absence of light until complete solubilization of formazan. The absorbance was read at 570 nm wavelength in a microplate spectrophotometer (Epoch2, BioTek, Winooski, VT, USA). Each condition was tested in 3 independent assays.

Nitric oxide was then measured to determine the anti-inflammatory potential of the extracts. For that, all the concentrations whose cell viability was above 90% were selected for the assay. A Griess diazotization reaction was used to measure the production of NO in RAW 264.7 cells according to Bahiense et al. with slight modifications [19]. Briefly, the microplates were seeded with 1×10^5 cells/well and incubated at 37 °C in 5% CO_2 for 24 h. After that period, cells were treated with the extracts for 6 h, following the addition of LPS (lipopolysaccharide) solution from *E. coli* (Sigma, Darmstadt, Germany) at a final concentration of 1.5 µg·mL^{-1} for 22 h. Then, 150 µL of the supernatants of the cell culture was mixed with 50 µL of Griess reagent (Sigma, Germany) and incubated for 15 min at room temperature. The absorbance was measured at 540 nm using a microplate spectrophotometer (Epoch2, BioTek, Winooski, VT, EUA). Each condition was tested using 6 technical replicates and 3 independent assays.

2.8. Data Treatment

The values of antioxidant activity were studied as specific activity (activity per mass unit of extract) and total activity (activity per unit of seaweed extracted), the latter calculated by multiplying the values of the respective activities by the yield of extract. All the graphs and statistical analysis were performed with GraphPad Prism v.6 (GraphPad Software, La Jolla, San Diego, CA, USA).

For ORAC and UV AUC activities, Holm–Sidak's multiple comparisons test was performed to understand the significant differences between extracts (different letters represent statistically significant differences, with $p < 0.05$).

For anti-enzymatic activity, cytotoxicity evaluation, photoprotection and anti-inflammatory potential, a one-way ANOVA was performed followed by Dunnett's multiple comparisons test to evaluate the significant differences between the extracts and the respective controls (* $p < 0.05$, ** $p < 0.01$, *** $p < 0.001$, **** $p < 0.0001$).

For antimicrobial activity, a two-way ANOVA followed by Tukey's multiple comparisons test was performed to evaluate the significant differences between extracts and between extracts and the inhibition control at each concentration (* $p < 0.05$, ** $p < 0.01$, m, **** $p < 0.0001$).

3. Results

A total of four hydroethanolic and two aqueous extracts, obtained from the biomass of *Grateloupia turuturu*, were selected according the previous study of Félix and co-workers [9]. Several bioactivities related to cosmetic/cosmeceutical applications were analyzed to understand the potential of these extracts in this field, specifically their antioxidant and UV absorbance capacity, and anti-enzymatic and antimicrobial activities, as well as their photoprotection and anti-inflammatory potential.

3.1. Antioxidant Activity and UV Absorbance

Two different concentrations of ethanol were used in the extraction procedure, generating hydroethanolic extracts with different compositions (Table 2). Regarding the yield obtained, it was possible to verify that in the presence of ethanol, yields were lower when compared with the aqueous extracts, reaching almost twice the percentage of the yield (minimum obtained for E4 with 23.50% and maximum for E6 with 50.84%). However, higher values of antioxidant activity using the ORAC method were found for the extracts with higher concentrations of ethanol, reaching, for E1, the maximum with 153.09 µmol of Trolox equivalents· $^{-1}$ extract and, for E6, the minimum with 45.00 µmol of Trolox equivalents·g^{-1} extract ($p < 0.05$). Similarly, E1 and E2 were the extracts presenting the highest values of UV absorbance.

3.2. Anti-Enzymatic Activity

Two different enzymes, known to be involved in skin degradation, were selected for this study. The inhibition of elastase (Figure 1A) and hyaluronidase (Figure 1B) activities was analyzed using the six seaweed extracts at 2 mg·mL^{-1}. Results show that for elastase, all the extracts were able to inhibit nearly 100% of enzymatic activity when compared with the control (Figure 1A). For hyaluronidase, the inhibition percentages were also above 77% for all extracts, except for extract 1 (E1), which presented the lowest value of inhibition for this enzymatic activity (close to 40% inhibition).

Table 2. Selected extracts of *Grateloupia turuturu* and respective ethanol percentage, yield, antioxidant capacity by ORAC assay (expressed as μmol of Trolox equivalents (TE) per gram of extract) and UV absorbance, using the area under the curve. Holm–Sidak's multiple comparisons test was performed to understand the significant differences between extracts (different letters represent statistically significant differences, with $p < 0.05$).

Extracts	EtOH (%)	Yield (%)	ORAC (μmol TE·$^{-1}$ ext)			UV AUC		
			Mean	SD	Significant Differences	Mean	SD	Significant Differences
E1	50	24.39	153.1	11.37	a	5.82	0.25	a
E2	50	28.56	102.3	8.33	b	4.06	0.16	b
E3	25	24.28	45.98	2.82	c	1.63	0.13	c
E4	25	23.50	66.81	6.79	d	3.20	0.03	d
E5	0	43.37	50.26	2.89	c	3.08	0.25	d
E6	0	50.84	45.00	3.77	c	2.26	0.30	e

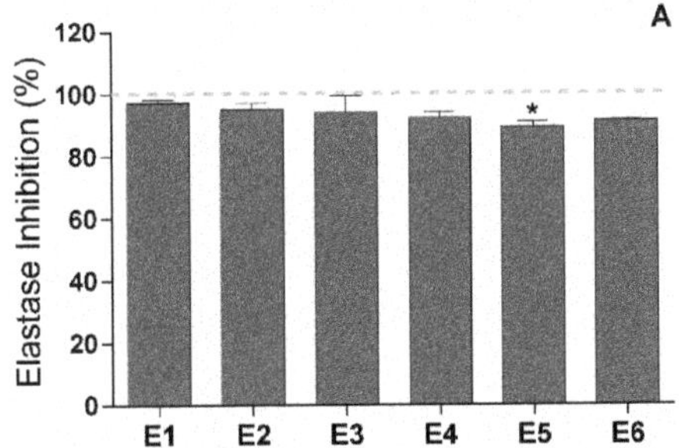
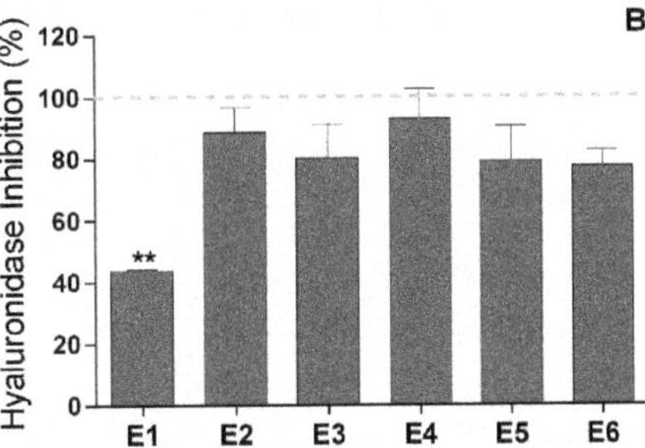

Figure 1. Evaluation of anti-enzymatic activity of the six extracts of *Grateloupia turuturu* at 2 mg·mL^{-1}: inhibition of elastase (**A**) and hyaluronidase (**B**) activities. Control of inhibition is represented as a dashed line. A one-way ANOVA followed by Dunnett's multiple comparisons test was performed to evaluate the significant differences between the extracts and the inhibition control (* $p < 0.05$, ** $p < 0.01$). Values presented are the mean of 3 independent assays.

3.3. Antimicrobial Activity

The antimicrobial potential of the extracts was evaluated against three representative microorganisms, namely, a Gram-negative bacterium, *Escherichia coli*, a Gram-positive bacterium, *Staphylococcus aureus*, and a fungal species, *Candida albicans* (Table 3). Regarding the bacterial inhibition of *E. coli*, extracts E1 and E2 (with higher percentages of ethanol used for the extraction procedure—50%) should be highlighted since the lower concentrations of extracts tested (0.0075 and 0.75 mg·mL^{-1}) were significantly different ($p < 0.05$ or less) from the same concentrations for the other extracts (Table S1—complete statistical analysis). However, for the other concentrations (1.5 and 3 mg·mL^{-1}), no significant differences were found between extracts, with the exception of E2 and E4 that significantly differ from each other ($p < 0.05$). The highest values of inhibition were found for E1 and E2, for 0.75 and 1.5 mg·mL^{-1}, reaching values near to 40% of inhibition.

Table 3. Antimicrobial activity of *Grateloupia turuturu* extracts against the bacteria *Staphylococcus aureus* and *E. coli* and the fungus *Candida albicans* at 4 different concentrations of extracts: 0.0075, 0.75, 1.5 and 3 mg·mL^{-1}. A two-way ANOVA followed by Tukey's multiple comparisons test was performed to evaluate the significant differences between extracts and between extracts and the inhibition control at each concentration (see Supplementary Table S1). Values presented are the mean of 3 independent assays.

						***Staphylococcus aureus***								
	E1		E2		E3		E4		E5		E6		C + (Tetracycline)	
mg·mL^{-1}	Mean	SD	Mean	SD	Mean	SD	Mean	SD	Mean	SD	Mean	SD	Mean	SD
3	17.22	5.25	37.77	5.84	54.01	1.48	28.37	4.50	32.40	7.95	35.80	4.68	100.56	0.26
1.5	15.14	4.36	26.88	5.85	52.99	3.67	25.60	4.41	23.72	6.71	24.90	7.99	101.18	1.23
0.75	29.58	8.93	35.13	8.60	52.64	3.66	38.33	2.88	10.79	21.08	4.28	23.15	100.54	0.22
0.0075	−8.93	2.33	−3.49	3.31	−9.23	5.29	−5.86	5.99	−10.58	6.05	−10.63	5.72	100.23	1.02

						Escherichia coli								
	E1		E2		E3		E4		E5		E6		C + (Ciprofloxacin)	
mg·mL−1	Mean	SD	Mean	SD	Mean	SD	Mean	SD	Mean	SD	Mean	SD	Mean	SD
3	21.14	1.70	25.79	0.17	16.67	3.28	11.33	4.88	17.58	5.80	13.87	6.57	106.10	2.57
1.5	30.69	0.72	32.99	2.01	25.68	7.04	23.10	9.54	27.07	10.52	26.18	10.79	106.10	2.57
0.75	33.89	4.43	36.27	5.31	10.51	5.23	7.00	3.97	4.61	2.92	15.73	4.40	105.69	0.71
0.0075	20.04	0.17	29.28	11.34	4.85	1.00	3.95	0.44	3.45	1.92	4.51	1.70	106.01	0.80

						Candida albicans								
	E1		E2		E3		E4		E5		E6		C + (Amphotericin B)	
mg·mL−1	Mean	SD	Mean	SD	Mean	SD	Mean	SD	Mean	SD	Mean	SD	Mean	SD
3	7.40	2.22	4.77	2.16	41.32	6.95	31.54	1.87	37.64	2.42	37.18	2.04	101.42	0.54
1.5	3.87	0.47	5.45	1.15	21.08	5.31	31.11	0.98	23.94	4.97	32.56	3.03	101.42	0.54
0.75	20.98	6.65	28.09	1.68	−26.55	12.72	−14.58	8.44	−8.71	4.34	−3.06	14.70	105.27	2.30
0.0075	23.89	2.22	8.69	6.14	41.53	4.44	55.26	2.44	11.45	4.48	55.39	0.54	103.90	3.62

In the case of *S. aureus* inhibition, E3 showed to be the most promising extract against this bacterium, reaching values of inhibition close to 60% between 0.75 and 3 mg·mL^{-1} (significantly different from the other extracts, $p < 0.05$ or less), and only the lowest concentration, 0.0075 mg·mL^{-1}, presented values of inhibition below 10%. It is also possible to verify that hydroethanolic extracts (E1–E4) were more efficient at inhibiting the growth of *S. aureus* (mostly at 0.75 and 3 mg·mL^{-1}) when compared with aqueous extracts (E5–E6), with E6 at 0.0075 mg·mL^{-1} being responsible for the opposite effect—bacterial growth promotion.

The antimicrobial activity of the extracts against *C. albicans* showed a more variable profile between extracts and concentrations when compared to bacteria. In fact, the highest values of inhibition correspond to the E4 and E6 extracts at 0.0075 mg·mL^{-1} ($p < 0.0001$) and the lowest values (fungal growth promotion) were found for the same extracts but at 0.75 mg·mL^{-1} ($p < 0.05$ or less). Except for those cases, E1 and E2 were the extracts that reached lower values of inhibition.

Comparing the ability of the extracts in the study against the three microorganisms, globally, the higher inhibition (near to 60%) was found against *S. aureus* (E3) and *C. albicans* (E4 and E6) with different extracts, while against *E. coli* were the extracts E1 and E2 that were responsible for the higher antibacterial inhibition (near to 40%).

3.4. Photoprotection Activity

A dose–response evaluation of each extract was performed in a fibroblast cell line, 3T3, using a range of concentrations between 0.01 and 4 mg·mL^{-1}, in order to evaluate the security of the extracts for skin applications (Figure S1). E1 and E2 (extraction with 50% ethanol/50% water) were the extracts with cell toxicity associated with more concentrations, especially E2, where cell viability was above 80% only in two of the eight concentrations tested. For E4, no concentration revealed a cytotoxic effect on 3T3 when compared with the control, and for E3 and E6, only the highest concentration (4 mg·mL^{-1}) was responsible for a reduction in cell viability of close to 80% ($p < 0.0001$), with all the other concentrations being above 80% of cell viability. E5 showed significant differences when compared to the control ($p < 0.0001$) for the concentrations of 2 and 4 mg·mL^{-1}, but also kept the values of cell viability near to 80%. Mostly in extracts with a lower or no concentration of ethanol in the extraction procedure (E3–E6), it was also possible to verify that the lowest concentrations tested were responsible for an increase in lysosomal activity, which might indicate growth promotion, being significantly different from the control.

Based on the results obtained for cytotoxicity in 3T3 cells, a photoprotection assay using the concentrations closer to 100% of cell viability found for each extract was performed (Figure 2). Cells were exposed to a UV radiation dose capable of killing 50% of cells in the presence and absence of extracts. The results showed that none of the extracts tested presented a photoprotection capacity. From the six extracts, E1, E3 and E6 did not show any differences when compared with the control (cells without extracts and exposed to UV radiation), while E2 ($p < 0.05$), E4 ($p < 0.0001$) and E5 ($p < 0.001$) revealed a phototoxic behavior.

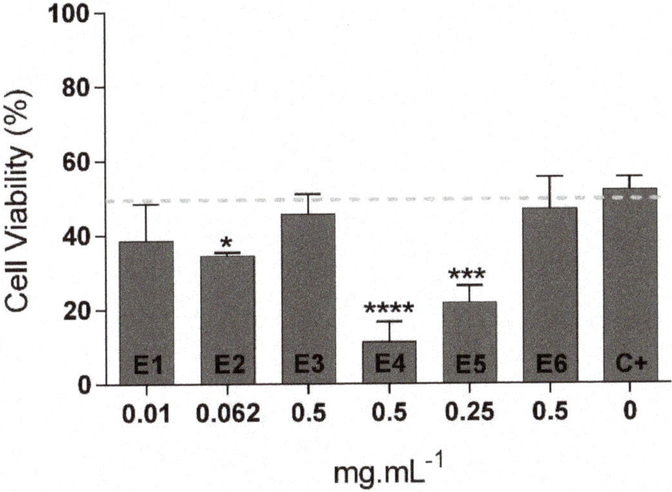

Figure 2. Photoprotection assay using the closest concentration to 100% of cell viability identified for each extract of *Grateloupia turuturu*. Cells were subjected to UV radiation in the presence and absence of extracts until IC$_{50}$ of control without extract was reached to evaluate the photoprotection potential. Control of cell viability is represented as a dashed line. A one-way ANOVA followed by Dunnett's multiple comparisons test was performed to evaluate the significant differences between the extracts and the control (* $p < 0.05$, *** $p < 0.001$, **** $p < 0.0001$). Values presented are the mean of 3 independent assays.

3.5. Nitric Oxide (NO) Measurement

A dose–response evaluation of each extract was performed in a macrophage cell line using the same range of concentrations used for 3T3 cells (Figure S2). Similar patterns were found in both cases: extracts with a higher concentration of ethanol in the extraction procedure (E1 and E2) showed higher cytotoxic effects when compared with the other four extracts. For the hydroethanolic extracts with 25% ethanol, the highest decrease in cell viability was reached for the concentration of 4 mg·mL^{-1} ($p < 0.0001$) with less than 20% of cell viability. The same trend was found for aqueous extracts, but although the concentration of 4 mg·mL^{-1} was significantly different from the control, cell viability percentages for that concentration were still high (above 70%).

For each extract, all the concentrations above 90% cell viability were used to analyze the nitric oxide production and consequently the anti-inflammatory potential (Figure 3).

Comparing with the control (cells subjected only to LPS solution), two extracts showed significant differences: E3 at 0.25 mg·mL^{-1} ($p < 0.05$), reducing the NO production 20%, and E6 at 0.01 ($p < 0.01$) and 0·25 mg.mL^{-1} ($p < 0.0001$), reducing 27% and 38.3%, respectively. However, higher concentrations increased the NO production on macrophages cells.

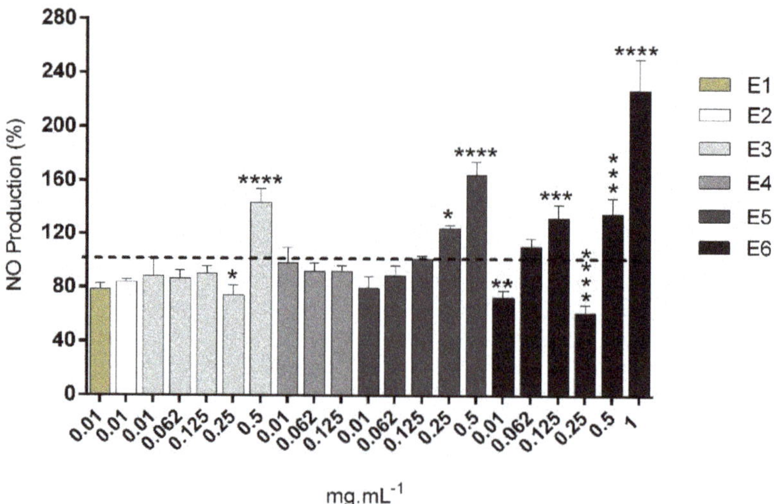

Figure 3. Nitric oxide assay was performed using all the non-cytotoxic concentrations identified for each extract of *Grateloupia turuturu* to evaluate their anti-inflammatory potential, using a final concentration of 1.5 µg·mL^{-1} of LPS. Control of cell viability is represented as a dashed line. A one-way ANOVA followed by Dunnett's multiple comparisons test was performed to evaluate the significant differences between the extracts and the control (* $p < 0.05$, ** $p < 0.01$, *** $p < 0.001$, **** $p < 0.0001$). Values presented are the mean of 3 independent assays.

4. Discussion

The introduction of natural ingredients in the cosmetic industry is continuously increasing in an attempt to find effective, safer and sustainable solutions. Therefore, the screening of bioactives from marine resources to apply in the cosmetic/cosmeceutical field is a great contribution to achieve that. Building upon the work of Felix et al. [9], where the effects of the percentage of ethanol, temperature, time, pH, and solid-to-liquid ratio were all characterized in the solid–liquid extraction of *Grateloupia turuturu's* antioxidant and UV-shielding compounds, six selected extracts were chosen and produced to further evaluate their properties of interest for the cosmeceutical industry.

The yield obtained for the six extracts (Table 1) showed that the increasing concentrations of ethanol are responsible for the decrease in the yield percentage (minimum obtained for E4 with 23.5% and maximum for E6 with 50.8%). This is in agreement with the fact that water is able to extract not only the galactans but also the proteins of this species, both presenting a significant massic contribution, while the presence of ethanol is responsible for their solubility decrease [9,20,21].

For the antioxidant activity by ORAC (Table 1), the opposite result was found: higher values of antioxidant activity in the presence of higher percentages of ethanol (reaching, for E1, the maximum with 153.09 and, for E6, the minimum with 45.00 µmol of Trolox equivalents·g^{-1} extract). This is also in accordance with the bibliography, since alcohols are known to be more efficient in the recovery process of antioxidants [20]. The presence of ROS, mostly originated from UV exposure, is responsible for triggering several processes in the skin (such as inflammation, oxidation of surface skin, hyperpigmentation and degradation of the dermal matrix, among others), promoting skin damages. Therefore, the use of molecules with antioxidant activity is a widely used approach to control and prevent symptoms related to skin damage [21].

Specifically, in *G. turuturu*, the presence of ethanol during the extraction procedure will contribute to recover molecules such as chlorophylls, polyphenols, and polar carotenoids and tocopherols, known for their antioxidant capacity [22–24], and thus, desired compounds to apply in cosmetic formulations. Although the main antioxidants are recovered using solvents such as ethanol, this does not prevent the study of aqueous extracts, since sulfated carrageenans, water-soluble compounds typically found in red algae, are also known for their antioxidant activity [25,26]. Apart from that, these polysaccharides are widely used in several industries due to their biocompatibility and high viscosity and gel forming properties [27].

Concerning the UV absorbance (Table 1), the highest ethanol concentration resulted in extracts with higher values of UV absorbance. This may be related to the extraction of compounds such as MAAs and polyphenols using this hydroethanolic mixture (50% each), while compounds with no activity, such as carbohydrates, are poorly extracted [9,28].

Between the different types of damages caused by the oxidative stress is the degradation of the extracellular matrix, which leads to a decrease in components responsible for the structure, elasticity and hydration of the skin (such as collagen, elastin and hyaluronic acid) and, consequently, to signs of skin aging, such as thinner skin, fine lines and wrinkles [2]. Thus, compounds able to enhance the inhibition of collagenase, elastase and hyaluronidase, among others, may be potential targets to use as bioactive ingredients in products with anti-aging properties [29]. In this context, the inhibition of elastase (Figure 1A) and hyaluronidase (Figure 1B) activities was analyzed using the six seaweed extracts at 2 mg·mL^{-1}. For elastase, all the extracts were able to inhibit near to 100% of enzymatic activity when compared with the control (Figure 1A), and for hyaluronidase, the inhibition percentages were also above 77% for all extracts (with the exception of E1), showing a great potential as active ingredients for anti-wrinkle formulations. The high percentages of inhibition for all extracts suggest that more than one type of compound is responsible for these bioactivities, since the presence of different concentrations of ethanol during the extraction, or even the absence, would result in the extraction of different classes of compounds. While sulfated polysaccharides and proteins are almost exclusively soluble in water, compounds such as carotenoids, sterols and fatty acids, among others, are preferentially extracted using a compromise between ethanol and water due to their medium polarity. For MAAs, it is expected that from water to higher percentages of ethanol, the extraction of these compounds would occur, possibly presenting different relative contents for each extraction condition [9]. Peptides from seaweed, such as signal peptides, were described to stimulate the extracellular matrix, increasing neocollagenesis and elastin synthesis, resulting in wrinkle reduction and skin firming [2,30]. Moreover, secondary metabolites, such as MAAs, have been described by their anti-wrinkle ability [29], mostly by their ability to inhibit the collagenase and elastase activities and to stimulate the secretion of hyaluronic acid by human fibroblasts [31–33]. Another group of secondary metabolites produced by red macroalgae, known for their capacity to maintain the extracellular matrix as healthier, are phenolic compounds [10]. In fact, a study conducted using a red macroalgae resulted in a methanolic extract rich in phenolic compounds, which was able to inhibit the overexpression of metalloproteinases, preventing the formation of wrinkles [31].

The antimicrobial properties of seaweed are also well established for a wide range of macroalgae [34,35]. They are known to produce bioactive compounds to inhibit/reduce the growth of other competitive microorganisms [34]. For the six hydroethanolic extracts, three microorganisms were selected to evaluate the antimicrobial activity of *G. turuturu*: a Gram-negative and a Gram-positive bacterium, *E. coli* and *S. aureus*, respectively, and a fungal strain of *C. albicans* (Table 3). Results show that for *E. coli*, extracts with higher percentages of ethanol were responsible for higher percentages of growth inhibition (E1 and E2, reaching near to 40% of inhibition). In the case of the Gram-positive *S. aureus*,

E3 was the most promising extract with almost 60% of growth inhibition for three of the four concentrations tested, while for *C. albicans*, the antimicrobial profile obtained was not so clear, with some concentrations of the extracts E3–E6 being among the most effective against this species. The recent study of Cardoso and co-workers [7] showed that ethanolic and polysaccharide extracts of *G. turuturu* presented antibacterial activity against *E. coli* and *S. aureus*, corroborating the results obtained and showing that the polysaccharides, such as carrageenan, present in this species, have antimicrobial activity. Further, the antifungal activity of this species was already confirmed for several species, as stated by Plouguerné and co-workers, who found extracts of *G. turuturu* highly active against five fungi species [36]. Regarding the hydroethanolic extracts of *G. turuturu* (E1–E4), bioactive compounds such as sulfated polysaccharides, phenolic compounds and carotenoids may be present and responsible for the antimicrobial activity, since they are known to alter the microbial cell permeability and to interfere with the membrane, leading to the loss of cellular integrity [34]. The wide antimicrobial activity of these extracts shows the potential for the cosmetic industry, as functional ingredients, but also as natural preservatives of cosmetic formulations, increasing the shelf-life of the product by reducing the microbial contamination [2,34].

In this study, a dose–response evaluation of each extract was performed in a mouse fibroblast cell line, 3T3, since these cells are one of the mains constituents of the skin, using a range of concentrations between 0.01 and 4 mg·mL^{-1}, in order to evaluate the security of the extracts for skincare applications (Figure S1). E1 and E2 (extraction with 50% ethanol/50% water) were the extracts with cell toxicity associated with more concentrations, while for E3 to E6, only the highest concentrations were responsible for a reduction in cell viability, the lowest concentrations tested being responsible for an increase in the neutral red signal, suggesting a growth promotion. The presented results show that for all the extracts tested, several concentrations were not cytotoxic to fibroblast cells, being an excellent preliminary result about their security for potential applications in skincare products. Another important feature of red seaweeds for the cosmetic industry is the production of bioactive compounds with photoprotection activity, able to protect the skin from damages such as sunburn, photo-aging, photo-dermatoses and skin cancer, among others [2,11]. The production of such compounds by macroalgae consists of ecophysiological strategies developed to avoid the deleterious effects of the constant exposure to UV radiation, through the absorption of UV radiation [11,37]. In fact, bioactive compounds able to absorb UV radiation were found to protect human fibroblasts from cell death and to retard the signs of aging induced by UV radiation [2,38]. Red macroalgae are known to produce a variety of compounds with this ability, such as phenolic compounds, pigments and MAAs. Between them, MAAs are known to be the most relevant for this function [2,37]. These secondary metabolites present high antioxidant and UV absorbing capacities, acting as excellent UV filters and thus having a great potential for the cosmetic industry as antioxidants and photoprotectors [2,29].

Based on the results obtained for cytotoxicity in 3T3 cells, a photoprotection assay using the concentrations closer to 100% of cell viability found for each extract was performed (Figure 2). Cells were exposed to a UV radiation dose capable of reaching the IC$_{50}$ of cells in the presence and absence of extracts. The results show that none of the extracts tested presented photoprotection capacity. From the six extracts, E1, E3, and E6 did not show any differences compared with the control (cells without extracts and exposed to UV radiation), while E2 ($p < 0.05$), E4 ($p < 0.0001$), and E5 ($p < 0.001$) revealed a phototoxic behavior, which was not an expected result for hydroethanolic extracts of the red seaweed *G. turuturu*. Comparing with the UV absorbance capacity of each extract, it is not possible to correlate the data, since E1 and E2, extracts with higher ethanol concentrations and consequently more phenolic compounds, were not the ones presenting less phototoxicity. However, several factors may contribute to explain these results. Specifically, a dose–response evaluation for the phototoxicity test could help to understand the effect of extract concentrations on this bioactivity. Since we are working with crude extracts that present a mixture of compounds with synergistic/antagonistic effects, it is possible that the best solution for the desired function may be related to specific concentrations where the dilution of certain compounds would be beneficial. Another interesting fact is that not all the MAAs have the same ability to act as photoprotectors [37]. At this point, further chemical characterization of the extracts could help to discriminate the presence of MAAs and which of them are present/at a relative quantity. Regarding also the production of MAAs, it is known that their production is affected by several abiotic factors, preferring the summer period and moderate depth [29,39]. A recent study also focused on the fact that *G. turuturu* tends to reduce the production of MAAs in the presence of intense UV radiation [40], suggesting that the production of such compounds may be highly influenced by different factors that are not totally controlled when the macroalga is grown in a natural environment.

The anti-inflammatory potential of seaweed has also been explored in an attempt to find potential sustainable and safer solutions with less side effects, especially for treatments of chronic inflammation [2]. During inflammation, oxidative stress increases and the cellular antioxidant capacity decreases, leading to large quantities of produced free radicals that will interact with fatty acids, cell membranes, proteins, and other components, promoting permanent alterations in cellular functions [41]. This process is mediated by a system of soluble factors that differ in their source and composition, one of them being the production of nitric oxide by macrophages. This compound is responsible for inducing vasodilatation, acting as a cytotoxic agent for pathogens [41]. Therefore, the discovery of novel compounds able to act as anti-inflammatories could be a new insight in this field.

A dose–response evaluation of each extract was performed in a macrophage cell line due to the direct implication of these cells in inflammatory processes, using the same range of concentrations used for 3T3 cells (Figure S2). Similar patterns were found in both cases: extracts with a higher concentration of ethanol in the extraction procedure (E1 and E2) showed higher cytotoxic effects when compared with the other four extracts. For each extract, all the concentrations above 90% cell viability were used to analyze the nitric oxide production (Figure 3). Comparing with the control (cells subjected only to LPS solution), two extracts showed significant differences: E3 at 0.25 mg·mL^{-1} ($p < 0.05$), reducing the NO production to 80%, and E6 at 0.01 ($p < 0.01$) and 0.25 mg·mL^{-1} ($p < 0.0001$), reducing 27% and 38.3%, respectively. However, higher concentrations stimulated NO production on macrophages cells.

In macroalgae, several types of bioactive compounds have already been described for their anti-inflammatory potential, namely, pigments (such as carotenoids), sulfated polysaccharides, proteins and their derivatives (such as phycobiliproteins), fatty acids (such as polyunsaturated fatty acids) and other compounds such as halogenated compounds or terpenes [10,29]. Some of them are known to be produced by red macroalgae and specifically by *G. turuturu*, such as carotenoids, phycobiliproteins, and sulfated polysaccharides, among others, which could help to explain the reduction in NO production in those extracts. However, different methods for anti-inflammatory evaluation, such as Western blot quantification of inflammatory markers' expression (e.g., TNF-α, interleukins, among others), should be implemented, since different pathways may be activated and the specificity of the technique is higher.

5. Conclusions

The potential of the invasive macroalga *G. turuturu* for the cosmetic industry was investigated. Several bioactivities concerning skin protection of the hydroethanolic extracts were analyzed and the results obtained show that different concentrations of ethanol led to extracts with different bioactivities. Noticeably, among the tested extracts, good antioxidant and antimicrobial activities were found, which promotes the added value of these extracts both for skin benefits and for formula's benefits. Additionally, significant inhibition of skin aging-related enzymes was attained, as well as some degree of inhibition of the inflammatory marker NO. A photoprotection assessment allowed the discovery of a phototoxicity of some extracts from *G. turuturu*, which is unexpected but very important information concerning this biomass use and valorization. This study is, therefore, an important contribution to understanding that seaweed extracts obtained from simple solvents (ethanol and water) and techniques (SLE), compatible with the industrial scale, have potential to be applied in the cosmetic field, bridging the demand for natural, greener and more sustainable products. However, further fractionation and/or characterization of these crude extracts is essential to understand the active ingredients of each extract responsible for the analyzed bioactivities.

Supplementary Materials: The following are available online at https://www.mdpi.com/2076-3417/11/4/1650/s1. Table S1: Statistical analysis of antimicrobial data using a two-Way ANOVA followed by Tukey's multiple comparisons test. Significant differences between extracts and between extracts and inhibition control at each concentration tested were analyzed and discriminated (* p < 0.05, ** p < 0.01, *** p < 0.001, **** p < 0.0001). Figure S1: Dose–response cytotoxic evaluation of the E1 (A), E2 (B), E3 (C), E4 (D), E5 (E) and E6 (F) extracts of *Grateloupia turuturu* at 8 different concentrations (0.01, 0.062, 0.125, 0.25, 0.5, 1, 2 and 4 mg·mL-1) in 3T3 cells. Figure S2: Dose–response cytotoxic evaluation of the E1 (A), E2 (B), E3 (C), E4 (D), E5 (E) and E6 (F) extracts of *Grateloupia turuturu* at 8 different concentrations (0.01, 0.062, 0.125, 0.25, 0.5, 1, 2 and 4 mg·mL^{-1}) in RAW 264.7 cells.

Author Contributions: Conceptualization, A.M.C., R.F., S.C.N. and M.F.L.L.; methodology, R.F.; formal analysis, R.F. and C.F.; investigation, C.F., A.M.C., A.J., P.D. and T.V.; resources, M.F.L.L. and R.P.; writing—original draft preparation, C.F.; writing—review and editing, C.F., R.F. and M.F.L.L.; supervision, J.S., C.A. and M.F.L.L.; project administration, M.F.L.L.; funding acquisition, M.F.L.L. All authors have read and agreed to the published version of the manuscript.

Funding: This study was supported by UID/MAR/04292/2020 with funding from FCT/MCTES through national funds, and by the grant awarded to Rafael Félix (SFRH/BD/139763/2018). The authors also wish to acknowledge the support of the European Union through the EASME Blue Labs project AMALIA, Algae-to-MArket Lab IdeAs (EASME/EMFF/2016/1.2.1.4/03/SI2.750419), project VALORMAR (Mobilizing R&TD Programs, Portugal 2020), co-funded by COMPETE (POCI-01-0247-FEDER-024517), the Integrated Programme of SR&TD "Smart Valorization of Endogenous Marine Biological Resources Under a Changing Climate" (reference Centro-01-0145-FEDER-000018) and RD&T co-promotion project ORCHESTRA (nº 70155), co-funded by Centro 2020 program, Portugal 2020, European Union, European Regional Development Fund, and SAICTPAC/0019/2015—LISBOA-01-0145-FEDER-016405 Oncologia de Precisão: Terapias e Tecnologias Inovadoras (POINT4PAC), through the European Regional Development Fund.

Institutional Review Board Statement: Not applicable.

Informed Consent Statement: Not applicable.

Data Availability Statement: Data is contained within the article or supplementary material.

Conflicts of Interest: The authors declare no conflict of interest. The funders had no role in the design of the study; in the collection, analyses, or interpretation of data; in the writing of the manuscript, or in the decision to publish the results.

References

1. Jesumani, V.; Du, H.; Aslam, M.; Pei, P.; Huang, N. Potential use of seaweed bioactive compounds in skincare—a review. *Mar. Drugs* **2019**, *17*, 688. [CrossRef]
2. Pimentel, F.B.; Alves, R.C.; Rodrigues, F.; Oliveira, M.B.P.P. Macroalgae-derived ingredients for cosmetic industry-an update. *Cosmetics* **2018**, *5*, 4–9. [CrossRef]
3. Freitas, C.; Araújo, R.; Bertocci, I. Patterns of benthic assemblages invaded and non-invaded by Grateloupia turuturu across rocky intertidal habitats. *J. Sea Res.* **2016**, *115*, 26–32. [CrossRef]
4. Sekar, S.; Chandramohan, M. Phycobiliproteins as a commodity: Trends in applied research, patents and commercialization. *J. Appl. Phycol.* **2008**, *20*, 113–136. [CrossRef]
5. Terasaki, M.; Narayan, B.; Kamogawa, H.; Nomura, M.; Stephen, N.M.; Kawagoe, C.; Hosokawa, M.; Miyashita, K. Carotenoid profile of edible japanese seaweeds: An improved hplc method for separation of major carotenoids. *J. Aquat. Food Prod. Technol.* **2012**, *21*, 468–479. [CrossRef]
6. Li, W.; Su, H.N.; Pu, Y.; Chen, J.; Liu, L.N.; Liu, Q.; Qin, S. Phycobiliproteins: Molecular structure, production, applications, and prospects. *Biotechnol. Adv.* **2019**, *37*, 340–353. [CrossRef]

7. Cardoso, I.; Cotas, J.; Rodrigues, A.; Ferreira, D.; Osório, N.; Pereira, L. Extraction and analysis of compounds with antibacterial potential from the red alga Grateloupia turuturu. *J. Mar. Sci. Eng.* **2019**, *7*, 220. [CrossRef]
8. Araújo, R.; Violante, J.; Pereira, R.; Abreu, H.; Arenas, F.; Sousa-Pinto, I. Distribution and population dynamics of the introduced seaweed Grateloupia turuturu (halymeniaceae, rhodophyta) along the Portuguese coast. *Phycologia* **2011**, *50*, 392–402. [CrossRef]
9. Félix, R.; Carmona, A.M.; Félix, C.; Novais, S.C.; Lemos, M.F.L. Industry-friendly hydroethanolic extraction protocols for grateloupia turuturu UV-shielding and antioxidant compounds. *Appl. Sci.* **2020**, *10*, 5304. [CrossRef]
10. Lourenço-Lopes, C.; Fraga-Corral, M.; Jimenez-Lopez, C.; Pereira, A.G.; Garcia-Oliveira, P.; Carpena, M.; Prieto, M.A.; Simal-Gandara, J. Metabolites from macroalgae and its applications in the cosmetic industry: A circular economy approach. *Resources* **2020**, *9*, 101. [CrossRef]
11. Ariede, M.B.; Candido, T.M.; Jacome, A.L.M.; Velasco, M.V.R.; de Carvalho, J.C.M.; Baby, A.R. Cosmetic attributes of algae - A review. *Algal Res.* **2017**, *25*, 483–487. [CrossRef]
12. Dávalos, A.; Gómez-Cordovés, C.; Bartolomé, B. Extending Applicability of the Oxygen Radical Absorbance Capacity (ORAC-Fluorescein) Assay. *J. Agric. Food Chem.* **2004**, *52*, 48–54. [CrossRef]
13. Madan, K.; Nanda, S. In-vitro evaluation of antioxidant, anti-elastase, anti-collagenase, anti-hyaluronidase activities of safranal and determination of its sun protection factor in skin photoaging. *Bioorg. Chem.* **2018**, *77*, 159–167. [CrossRef] [PubMed]
14. Adamczyk, K.; Olech, M.; Abramek, J.; Pietrzak, W.; Kuźniewski, R.; Bogucka-Kocka, A.; Nowak, R.; Ptaszyńska, A.A.; Rapacka-Gackowska, A.; Skalski, T.; et al. Eleutherococcus species cultivated in Europe: A new source of compounds with antiacetyl-cholinesterase, antihyaluronidase, anti-DPPH, and cytotoxic activities. *Oxid. Med. Cell. Longev.* **2019**, *2019*. [CrossRef] [PubMed]
15. CLSI. Methods for Dilution Antimicrobial Susceptibility Tests for Bacteria That Grow Aerobically. In *Clinical and Laboratory Standards Institute*; Approved Standard - M7-A7; Clinical and Laboratory Standards Institute: Wayne, PA, USA, 2006; Volume 26, ISBN 1562386255.
16. CLSI. *Reference Method for Broth Dilution Antifungal Susceptibility Testing of Yeasts*; Approved Standard - M27-A3; Clinical and Laboratory Standards Institute: Wayne, PA, USA, 2006; Volume 28, pp. 0–13.
17. Repetto, G.; del Peso, A.; Zurita, J.L. Neutral red uptake assay for the estimation of cell viability/cytotoxicity. *Nat. Protoc.* **2008**, *3*, 1125–1131. [CrossRef] [PubMed]
18. OECD. OECD Test Guideline 432: In Vitro 3T3 NRU Phototoxicity Test. 2004, pp. 1–15. Available online: https://www.oecd-ilibrary.org/environment/test-no-432-in-vitro-3t3-nru-phototoxicity-test_9789264071162-en (accessed on 12 February 2021).
19. Bahiense, J.B.; Marques, F.M.; Figueira, M.M.; Vargas, T.S.; Kondratyuk, T.P.; Endringer, D.C.; Scherer, R.; Fronza, M. Potential anti-inflammatory, antioxidant and antimicrobial activities of Sambucus australis. *Pharm. Biol.* **2017**, *55*, 991–997. [CrossRef] [PubMed]
20. Sultana, B.; Anwar, F.; Ashraf, M. Effect of extraction solvent/technique on the antioxidant activity of selected medicinal plant extracts. *Molecules* **2009**, *14*, 2167–2180. [CrossRef]
21. Masaki, H. Role of antioxidants in the skin: Anti-aging effects. *J. Dermatol. Sci.* **2010**, *58*, 85–90. [CrossRef] [PubMed]
22. Tiwari, B.K.; Troy, D.J. Seaweed Sustainability - Food and Non-Food Applications. Tiwari, B.K., Troy, D.J., Eds.; Elsevier Inc.: New York, NY, USA, 2015; ISBN 9780124186972.
23. Shalaby, E.A. Algae as promising organisms for environment and health. *Plant Signal. Behav.* **2011**, *6*, 1338–1350. [CrossRef]
24. Garcia-Vaquero, M.; Rajauria, G.; O'Doherty, J.V.; Sweeney, T. Polysaccharides from macroalgae: Recent advances, innovative technologies and challenges in extraction and purification. *Food Res. Int.* **2017**, *99*, 1011–1020. [CrossRef]
25. Ye, D.; Jiang, Z.; Zheng, F.; Wang, H.; Zhang, Y.; Gao, F.; Chen, P.; Chen, Y.; Shi, G. Optimized extraction of polysaccharides from Grateloupia livida (Harv.) yamada and biological activities. *Molecules* **2015**, *20*, 16817–16832. [CrossRef]
26. Tang, L.; Chen, Y.; Jiang, Z.; Zhong, S.; Chen, W.; Zheng, F.; Shi, G. Purification, partial characterization and bioactivity of sulfated polysaccharides from Grateloupia livida. *Int. J. Biol. Macromol.* **2017**, *94*, 642–652. [CrossRef] [PubMed]
27. Pacheco-Quito, E.M.; Ruiz-Caro, R.; Veiga, M.D. Carrageenan: Drug Delivery Systems and Other Biomedical Applications. *Mar. Drugs* **2020**, *18*, 583. [CrossRef] [PubMed]
28. Athukorala, Y.; Lee, K.; Song, C.; Ahn, C.; Shin, T.; Cha, Y.-J.; Shahid, F.; Jeon, Y.-J. Potential antioxidant activity of marine red alga grateloupia filicina extracts. *J. Food Lipids* **2003**, *10*, 251–265. [CrossRef]
29. Pereira, L. Seaweeds as source of bioactive substances and skin care therapy-Cosmeceuticals, algotheraphy, and thalassotherapy. *Cosmetics* **2018**, *5*. [CrossRef]
30. Malerich, S.; Berson, D. Next generation cosmeceuticals. The latest in peptides, growth factors, cytokines, and stem cells. *Dermatol. Clin.* **2014**, *32*, 13–21. [CrossRef] [PubMed]
31. Ryu, B.M.; Qian, Z.J.; Kim, M.M.; Nam, K.W.; Kim, S.K. Anti-photoaging activity and inhibition of matrix metalloproteinase (MMP) by marine red alga, Corallina pilulifera methanol extract. *Radiat. Phys. Chem.* **2009**, *78*, 98–105. [CrossRef]
32. Terazawa, S.; Nakano, M.; Yamamoto, A.; Imokawa, G. Mycosporine-like amino acids stimulate hyaluronan secretion by up-regulating hyaluronan synthase 2 via activation of the p38/MSK1/CREB/c-Fos/AP-1 axis. *J. Biol. Chem.* **2020**, *295*, 7274–7288. [CrossRef]
33. Orfanoudaki, M.; Hartmann, A.; Alilou, M.; Gelbrich, T.; Planchenault, P.; Derbré, S.; Schinkovitz, A.; Richomme, P.; Hensel, A.; Ganzera, M. Absolute configuration of mycosporine-like amino acids, their wound healing properties and in vitro anti-aging effects. *Mar. Drugs* **2020**, *18*, 35. [CrossRef] [PubMed]
34. Pérez, M.J.; Falqué, E.; Domínguez, H. Antimicrobial action of compounds from marine seaweed. *Mar. Drugs* **2016**, *14*, 52. [CrossRef]

35. Silva, A.; Silva, S.A.; Carpena, M.; Garcia-Oliveira, P.; Gullón, P.; Barroso, M.F.; Prieto, M.A.; Simal-Gandara, J. Macroalgae as a source of valuable antimicrobial compounds: Extraction and applications. *Antibiotics* **2020**, *9*, 642. [CrossRef] [PubMed]
36. Plouguerné, E.; Hellio, C.; Deslandes, E.; Véron, B.; Stiger-Pouvreau, V. Anti-microfouling activities in extracts of two invasive algae: Grateloupia turuturu and Sargassum muticum. *Bot. Mar.* **2008**, *51*, 202–208. [CrossRef]
37. Álvarez-Gómez, F.; Korbee, N.; Casas-Arrojo, V.; Abdala-Díaz, R.T.; Figueroa, F.L. UV photoprotection, cytotoxicity and immunology capacity of red algae extracts. *Molecules* **2019**, *24*, 341. [CrossRef]
38. Bedoux, G.; Hardouin, K.; Burlot, A.S.; Bourgougnon, N. *Bioactive components from seaweeds: Cosmetic applications and future development*; Elsevier: New York, NY, USA, 2014; Volume 71, ISBN 9780124080621.
39. Pereira, L. Seaweed flora of the european north atlantic and mediterranean. In *Springer Handbook of Marine Biotechnology*; Springer: Cham, Switzerland, 2015; pp. 65–178. ISBN 9783642539718.
40. de Ramos, B.; da Costa, G.B.; Ramlov, F.; Maraschin, M.; Horta, P.A.; Figueroa, F.L.; Korbee, N.; Bonomi-Barufi, J. Ecophysiological implications of UV radiation in the interspecific interaction of Pyropia acanthophora and Grateloupia turuturu (Rhodophyta). *Mar. Environ. Res.* **2019**, *144*, 36–45. [CrossRef] [PubMed]
41. Fernando, I.P.S.; Nah, J.W.; Jeon, Y.J. Potential anti-inflammatory natural products from marine algae. *Environ. Toxicol. Pharmacol.* **2016**, *48*, 22–30. [CrossRef]

Article

Calliblepharis jubata Cultivation Potential—A Comparative Study between Controlled and Semi-Controlled Aquaculture

Glacio Souza Araujo [1] , **João Cotas** [2] , **Tiago Morais** [3], **Adriana Leandro** [2], **Sara García-Poza** [2], **Ana M. M. Gonçalves** [2,4] **and Leonel Pereira** [2,*]

1   Ceará Federal Institute/IFCE-Campus Aracati, Aracati 62800-000, Brazil; glacio@ifce.edu.br
2   Department of Life Sciences, MARE-Marine and Environmental Sciences Centre, University of Coimbra, 3001-456 Coimbra, Portugal; jcotas@gmail.com (J.C.); adriana.leandro@uc.pt (A.L.); sara.g.poza@uc.pt (S.G.-P.); amgoncalves@uc.pt (A.M.M.G.)
3   Lusalgae, Lda., Incubadora de Empresas da Figueira da Foz, Rua das Acácias N° 40-A, 3090-380 Figueira da Foz, Portugal; tsmorais@lusalgae.pt
4   Department of Biology and CESAM, University of Aveiro, 3810-193 Aveiro, Portugal
*   Correspondence: leonel.pereira@uc.pt

Received: 29 September 2020; Accepted: 20 October 2020; Published: 27 October 2020

Abstract: *Calliblepharis jubata* is an edible red seaweed and a carrageenan primary producer, considered native in Figueira da Foz (Portugal). *C. jubata* has the particularity of producing only one kind of carrageenan, the iota fraction. However, this seaweed is not yet valuable for the food industry or even for human consumption. In this work, we characterize important biochemical compounds of *C. jubata* growing up within different cultivation techniques and wild specimens. The aim of this work is to know if there are differences between the biological compounds of interest and identify the advantages for human consumption and the food industry. The results supported the nutritional value of the seaweed, where the ones from inshore cultivation (T) were more identical to the wild specimens (F), than the indoor *C. jubata* (A, B, C). The parameters analyzed were fatty acids, carbohydrates and carrageenan content.

Keywords: *Calliblepharis jubata*; aquaculture; fatty acids; carbohydrates; carrageenan

1. Introduction

Several eastern countries traditionally consume seaweed as food, mainly Japan, China and Korea [1], due to its high nutritional value as a source of proteins, carbohydrates, vitamins and minerals [2,3]. The great economic interest is also justified by the growing demand for phycocolloid for different uses in the pharmaceutical, food and cosmetics industries [4–9], which has led several countries to cultivate seaweed [10]. The main hydrocolloids of commercial interest in marine algae (agar, carrageenan and alginate) are also known to have several biological activities, such as anticoagulant [11], antithrombotic [12], antiviral [13], immunostimulants [14], thus having great biotechnological applicability [15]. The variety of seaweed commercial applications discussed raises an important sustainable question, which cannot be taken lightly further. In order to sustain a steady supply, a solution that allows a minimal interaction with marine ecosystems is needed. Aquaculture can be that solution [16,17]. Food safety can only be achieved with a feasible and controllable food source. The seaweed aquaculture is considered to be one of the major hypothesis to be part of the solution to obtain global food security [5]. However to be part of the solution, the potential of seaweeds needs to be characterized and, then selected the seaweeds that are simple to be cultivated, without major problems. And also, the selected seaweeds need to have a good nutritional value as demonstrated by Leandro et al. [5].

Seaweed aquaculture is, above all, a production method which can be scaled up to very high proportions, maintaining several advantages [16,18]. Thus, aquaculture technologies are considered a sustainable alternative to optimizing outputs (quantity and quality) and reducing costs and pollution. These technologies and methods need to be adapted to the environmental request from the aquaculture location and species cultivated, because the aquaculture can be offshore or onshore and abiotic and biotic factors that influence the aquaculture system are very different, which will have a high influence in the aquaculture productivity [19].

Inland saline aquaculture is defined as land-based aquaculture using saline or saltpans groundwater. Inland saline water can differ in chemistry compared with coastal seawater and adjusting the chemistry or choosing species that are tolerant to the differences is one of the major challenges for expansion of inland saline aquaculture. For seaweed aquaculture, a sure water supply and disposal through terminal evaporation ponds provide critical and expensive infrastructure [20]. Therefore, inland saline aquaculture encompasses a number of culture species, systems and water types. The potential for expansion of these production systems is almost unlimited. With the increasing demands on potable water and marine coastal locations, use (or re-use) of inland saline waters provides a critical resource for high-quality seafood production using unproductive or even detrimental, resources. Integration of seaweed aquaculture in conventional fish farms, halophytic culture, would further increase the efficiency and sustainability of these food production systems [20].

All in all, basic and applied research into practical management systems for these systems is rapidly providing us with the knowledge of how to turn these into profitable farming ventures and novel food production methods are needed to further improve global wellbeing and inland saline aquaculture is bound to be a most valuable tool [20].

The future development work of aquaculture techniques, both in shore and off shore, has to account for studies into those subjects and relate those with designs prepared for strong water movement and rich inorganic nutrient concentrations in order to enhance nutrient uptake (off shore aquaculture) or with aeration, dividers for agitation and CO_2 adjustments for pH control (in shore aquaculture).

The benthic seaweed *Calliblepharis jubata* [21] (Figure 1), which belongs to Class Rhodophyta, Order Gigartinales and Family Cystocloniaceae, is a candidate with an interesting profile for cultivation and extraction of phycocolloids. It presents a dark reddish-brown, cartilaginous but flaccid stem, up to 30 cm in length. Stipe ± cylindrical, up to 10 cm in length; slightly branched rhizoid. Lanceolate blade, up to 15 cm wide, simple, irregularly pinned or dichotomously divided, coated by rolled or hook-shaped proliferations, similar to tendons up to 10 cm or more; blade surface with small spiny growths in well-developed specimens. They have rocks as habitat or are epiphyte, from the medium horizon of the intertidal level to the shallow infralittoral [22].

The studies in this seaweed are mainly about their polysaccharide content. Zinoun et al. [23] reported sulfohydrolase catalyze reaction in this species, which converts carrageenan precursors molecules (μ- and ν-carrabiose) into 3,6 anhydrogalactose derivatives modifying the non-gelling precursors characteristic into gelling carrageenans derivatives, such iota-carrageenan [24]. Deslandes et al. [25] recurred to spectroscopic techniques to elucidate about the monosacharides composition of four red seaweeds, which demonstrated that the main precursor of carrageenan in *C. jubata* is the ν-carrabiose. This seaweed carrageenan has been characterized by vibrational spectroscopy (FTIR-Raman and FITR-ATR), that gives a hybrid iota or iota-kappa carrageenan, with very low content of kappa-carrageenan, in the all *C. jubata* life cycle phase [26].

Figure 1. *Calliblepharis jubata.*

Thus, *C. jubata* is an iota-carrageenan primary producer and is considered native in Figueira da Foz (Coimbra, Portugal). However, this seaweed is not yet valuable for the food industry or even for human consumption [27].

In this work, we characterize important biochemical compounds (fatty acids, carbohydrates) and yield of an important compound (carrageenan) in wild specimens of *C. jubata* and, furthermore with different cultivation techniques to confirm if this species can be cultivated and be a new source of an important compound in the food industry and also, in human direct consumption. Still, there is a lack of information in literature about this species, with a single article of Zinoun and Cosson [28] supporting the possibility of *C. jubata*'s cultivation.

2. Materials and Methods

2.1. Reagents

Methanol was purchased from José Manuel Gomes dos Santos, Lda., Odivelas, Portugal and acetone from the Ceamed Lda., Funchal, Portugal. Ethanol was obtained from Valente e Ribeiro. Lda., Belas, Portugal and sodium hydroxide from Sigma-Aldrich GmbH, Steinheim, Germany.

2.2. Seaweed Collection

The specimens of *Calliblepharis jubata* were collected in Praia da Tamargueira, Buarcos, Figueira da Foz (40°10′18.6″ N, 8°53′44.4″ W), Portugal (Figure 2). Sampling was conducted in May 2020 from the sites with well-established *C. jubata* patches and without epiphytes or degradation visible at eyesight. The *C. jubata* specimens were collected in two different tidal pools. The pools distance was proximate 1 m between them horizontally to guarantee the most identical composition and physical state between the seaweed collected for analysis and cultivation. Once harvested, samples were stored in plastic bags for transport to the laboratory, in a cool box. The samples were then transported in cold boxes to the laboratory, where they were washed with filtered seawater in order to remove sands, contaminants and other organisms that they may have had. After that, the samples were separated:

some specimens were weighted and placed in culture systems, while the remaining biomass was rapidly washed with distilled water, cleaned with paper and stored at −20 °C for further analysis.

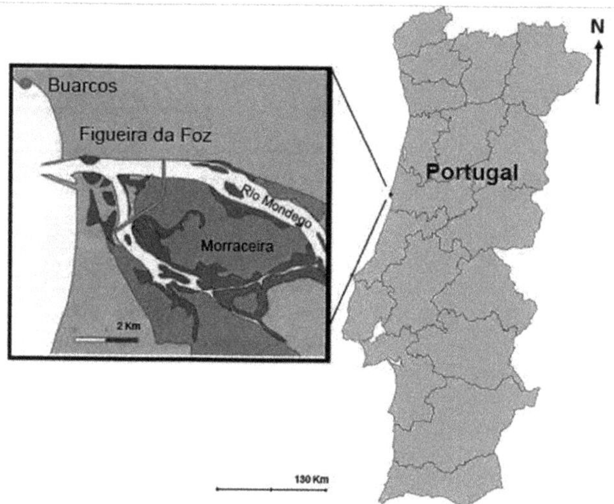

Figure 2. Sampling site: Praia da Tamargueira, Buarcos, Figueira da Foz, Coimbra, Portugal.

2.3. Seaweed Cultivation

Juveniles' specimens were selected during the primary seaweed collection and transported to the laboratory in plastic bags containing seawater and packed in cool boxes. Seawater was also collected to help in the specimens washing process, as well as to be the medium of the indoor cultivation first stage of the *C. jubata*, after filtering with paper filters.

2.3.1. Indoor Cultivation

Most of the juveniles used for cultivation (<5 cm) were selected for the indoor cultivation under controlled conditions at the Marine Algae Laboratory (MARE UC), Department of Life Sciences, Faculty of Science and Technology, University of Coimbra, Coimbra, Portugal.

The cultivation apparatus were round volumetric flasks of 1 L, containing 1 L of filtered seawater, with artificial light and artificial aeration from a diaphragm pump, 24 h per day of light. The light was on during periods of 16 h (day period 16 h^{L}:8 h^{D}), the light bar was at a distance of 10 cm of the flasks horizontally, the light was an OSRAM light tube with 18 W, 3100 LM and 4000 k.

The cultivation was carried out in triplicate, with average initial weights of 1.75 g ± 0.59 of *C. jubata* per liter of seawater, for a period of 42 days. The room temperature was 24 °C ± 2 °C and 55% ± 5% humidity. Every two days, the water in each container was renewed three times a week. Once a week, the algae were weighed on a semi-analytical balance to monitor their average growth.

2.3.2. Inshore Cultivation

In order to obtain more information, we recurred during the MENU—Marine macroalgaE—alternative recipes for a daily Nutritional diet project (Project Reference: FA_05_2017_011) to a one prototype cultivation tank from Lusalgae Lda. (Figueira da Foz, Coimbra, Portugal), to check the potential of *C. jubata* in the method applied in this seaweed aquaculture company. To obtain sufficient biomass for the extraction of compounds of biological interest and to compare them with seaweed grown under controlled conditions, in this stage, *C. jubata* grew in a water tank exposed to direct sunlight with aeration during the day (the aeration did not work during the night, only worked in the sunlight).

The specimens cultivated where identical to the Indoor cultivation (collected in the same pools in the same collection date, with length lower than 5 cm), to prevent the biochemical composition differential due to the seaweed growth state. With this, the cultivation method was the principal factor of differentiation of the biochemical profile.

The medium of cultivation was estuarine seawater (29%–34%) of the Mondego River estuary, in Figueira da Foz, Coimbra, Portugal, without adding nutrients.

The cultivation tank had a capacity of 1000 L, containing 800 L of mechanically filtered estuarine water.

The cultivation started from an initial biomass of 666 g in a single tank. Approximately 75% of the water volume was renewed three times a week and after three weeks the entire biomass was retired to further analysis. The seaweed cultivation method was done by the Lusalgae cultivation standardized method to the red seaweeds.

2.4. Seaweed Treatment before the Biochemical Characterization

The specimens were washed briefly with distilled water to remove salts and dried in an air force oven (Raypa DAF-135, R. Espinar S.L., Barcelona, Spain) at 60 °C, 48 h. The dried algae were finely ground with a commercial mill (Taurus aromatic, Oliana, Spain) ($\leq$1 mm) in order to render the samples uniform and then, stored in a dark room, in a box with silica gel to reduce the humidity, at ambient temperature ($\pm$24 °C). The dried biomass was stored to protect it from the light, until the analysis proceeds.

2.4.1. Fatty Acids Characterization

Samples were divided into three replicates and were processed in order to obtain the extracts. The lipids were extracted and methylated to fatty acids methyl esters (FAMEs) following the methodology described in Gonçalves et al. [29]. Samples were incubated with methanol for the extraction of lipids. The internal standard nonadecanoic acid C19 (Fluka 74208) was added to each sample to quantify the fatty acid (FA). Then Sodium chloride was added and the samples were centrifuged to the separation of FAMEs. Then, samples were centrifuged and stored at -80 °C. FAMEs identification was carried out through Gas chromatography-Mass spectrometry (GC-MS), equipped with a 0.32 mm internal diameter, 0.25 µm film thickness and 30 m long TR-FFAP column. The sample (1.00 µL) was injected with a splitless mode. The column temperature was programmed to increase from 80 to 230 °C, with helium as the carrier gas, at a flow rate of 1.4 mL min^{-1}. Integration of FAMEs peaks was carried out using the equipment's software. The identification of each peak was performed by retention time and mass spectrum of each FAME, comparing to the Supelco®37 component FAME mix (Sigma-Aldrich, Steinheim, Germany). Quantification of FAMEs was done as previously described in Gonçalves et al. [29].

2.4.2. Carbohydrates Content Characterization

The profiles of three replicates were extracted throughout a hydrolysis, followed by reduction and acetylation [30]. Samples were incubated with sulphuric acid and then added distilled water. After a reduction with ammonia and sodium borohydride, acetylation was performed with acid acetic, methylimidazole and acetic acid anhydride as described in Coimbra et al. [22]. The internal standard 2-desoxiglucose was added to the quantification of sugars. Samples were centrifuged and resuspended in acetone to be injected in the Gas-chromatography with a flame ionization detector (GC-FID). The samples were run through a Thermo Scientific Trace 1310 chromatography equipment (Waltham, MA, USA) equipped with a GC-FID. A TG-WAXMS A (30 m length, 0.32 mm i.d., 0.25 µm film thickness) GC column was used and the oven was programmed to an initial temperature of 180 °C, following a linear temperature increase of 5 °C min^{-1} until the final temperature of 230 °C was reached, maintaining this temperature for 12 min. The carrier gas was helium at a flow rate of 2.5 mL min^{-1}. The monosaccharides were identified by retention time comparison with standards. Quantification of sugars was performed by comparison of the sugar chromatographic peaks to the peaks obtained for the internal standard used (2-desoxiglucose).

2.5. Carrageenan Refined Alkali Extraction

Alkali extraction was performed according to the method described by Pereira et al. [31]. The milled seaweed was weighted in a scale (Radwag WLC 1/A2, Radwag, Radom, Poland) and 1 g samples was used ($n = 3$). Before extraction, the milled seaweed material (1 g) was resuspended and pretreated with an acetone: methanol (1:1) solution in a final concentration of 1% (m/v) for 16 h, at 4 °C, to eliminate the organic-soluble fraction. The liquid solution was decanted, and the seaweed residues obtained were dried in an air force oven (Raypa DAF-135, R. Espinar S.L., Barcelona, Spain) at 60 °C before the extraction method.

The samples were placed in 150 mL of NaOH (1 M) (1 g of initial seaweed: 150 mL of NaOH solution) in a hot water bath system (GFL 1003, GFL, Burgwedel, Germany), at 85–90 °C, for 3 h. The solutions were hot filtered, under vacuum, through a cloth filter supported in a Buchner funnel and a Kitasato flask. After that, the extracts were again filtered under vacuum with a Goosh 2 silica funnel. The extract was evaporated (rotary evaporator model: 2600000, Witeg, Germany) under vacuum to one-third of the initial volume. The carrageenan was precipitated by adding the warm solution to twice its volume of 96% ethanol. The carrageenan precipitated was washed with ethanol, 48 h at 4 °C before drying in an air force oven.

2.6. Statistical Analysis

For the characterization assays, the statistical analysis of variance (ANOVA) was performed, one for each component, followed by Tukey's tests in order to compare the values in the quantifications within the treatment (type of cultivation's tank or field samples). To determine whether any of the differences between the means were statistically significant, the p-value was compared to a significance level (denoted as α or alpha) of 0.05. Differences were considered significant when the p value was lower than 0.05.

3. Results

In the two seaweeds cultivation systems, we used identical specimens, collected in the same spots in the same collection date. However, at the end of the cultivation of *C. Jubata* under controlled conditions (42 days), a final average weight of 4.25 ± 1.09 g was obtained (Table 1). In the inshore aquaculture the initial biomass only grew 100 g, performing 766 g after three weeks.

Table 1. Average weight of *C. jubata* macroalgae during cultivation under controlled conditions. The values indicate the average weight of the macro algae ± standard deviation.

Days of Culture	Weight Average (g)
00	1.75 ± 0.59
07	2.28 ± 0.16
14	2.38 ± 0.11
21	2.78 ± 0.24
28	3.32 ± 0.41
35	4.00 ± 0.92
42	4.25 ± 1.09

3.1. Fatty Acids Characterization

Field and Tank cultivated samples were richer in total fatty acids than the cultivated samples under the controlled laboratory conditions. Table 2 shows the Tank T was richer in saturated fatty acids (SFA, 2186.69 µg·g^{-1} dw), while aquarium C was the poorest (1257.64 µg·g^{-1} dw). Furthermore, tank T was also the highest in monounsaturated fatty acids (MUFA, 305.74 µg·g^{-1} dw) and polyunsaturated fatty acids (PUFA, 137.96 µg·g^{-1} dw) concentration respectively, whereas aquarium B presented the lowest concentration on MUFA (266.03 µg·g^{-1} dw) and PUFA (20.79 µg·g^{-1} dw). However, the samples of the field had the highest concentration of highly unsaturated fatty acids (HUFA, 2602.41 µg·g^{-1} dw) and the lowest was aquarium B (506.73 µg·g^{-1} dw). Referring to the total fatty acid composition, field samples presented the highest value (5289.05 µg·g^{-1} dw) while aquarium B showed the lowest (2105.43 µg·g^{-1} dw) content (Figure 3).

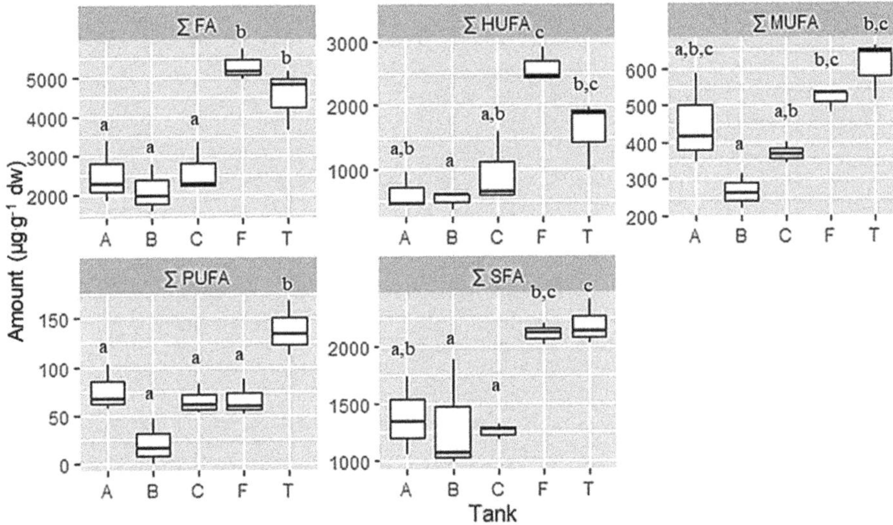

Figure 3. Sum of fatty acid group and total fatty acid expressed in µg·g^{-1} dw from samples of *C. jubata* obtained from cultivated in indoor balloons (A, B and C), cultivated in outdoor tank (T) and collected from the field (F). The same upper letters (a, b, c) mean there are no differences between the treatments (cultivation type). Data as mean value, distribution and variance; *n* = 3.

Table 2. Fatty acid composition and summary of fatty acid grouped in saturated (SFA), monounsaturated (MUFA), polyunsaturated (PUFA), highly unsaturated (HUFA) and total of fatty acids (FA). Results were expressed in $\mu g \cdot g^{-1}$ of dry weight (dw) from samples of *C. jubata* obtained from cultivated in indoor balloons (A, B and C), cultivated in outdoor tank (T) and collected from the field (F). Data as mean ± SEM; $n = 3$.

	A			B			C			F			T		
C14:0	244.83	±	34.54	209.08	±	51.07	233.02	±	11.43	343.22	±	9.74	353.59	±	20.52
C16:0	1013.99	±	144.06	1055.26	±	217.31	964.31	±	38.88	1707.55	±	44.05	1723.46	±	100.96
C18:0	116.05	±	24.74	47.53	±	15.68	60.31	±	6.42	53.99	±	4.71	109.64	±	9.11
Σ SFA	1374.88	±	200.39	1311.87	±	283.84	1257.64	±	40.01	2104.76	±	54.10	2186.69	±	113.37
C16:1	122.91	±	22.26	39.32	±	2.68	96.65	±	12.43	102.91	±	4.26	116.98	±	11.89
C18:1	325.12	±	50.33	226.71	±	25.59	272.73	±	16.77	413.02	±	19.16	488.77	±	43.88
Σ MUFA	448.03	±	71.97	266.03	±	26.73	369.38	±	15.95	515.92	±	17.57	605.74	±	46.09
C18:2	44.44	±	9.96	10.42	±	5.48	36.59	±	4.66	16.06	±	0.93	97.52	±	16.11
C18:3	31.01	±	4.38	10.38	±	8.94	28.10	±	4.60	49.90	±	10.39	40.44	±	7.48
Σ PUFA	75.44	±	13.89	20.79	±	13.73	64.69	±	8.80	65.95	±	10.98	137.96	±	16.42
C20:4	271.02	±	62.58	254.10	±	50.56	260.21	±	21.41	527.89	±	16.26	427.78	±	86.39
C20:5	331.31	±	109.17	252.63	±	34.40	659.80	±	337.32	2074.52	±	152.02	1171.47	±	242.51
Σ HUFA	602.33	±	171.38	506.73	±	74.56	920.01	±	335.24	2602.41	±	158.41	1599.24	±	315.65
Σ FA	2500.67	±	451.90	2105.43	±	347.98	2611.73	±	368.52	5289.05	±	228.40	4529.64	±	452.94

3.2. Carbohydrate Characterization

Monosaccharides present in the samples were analysed after hydrolysis of the polysaccharides. Our method does not permit to separate the two hexoses, glucose and galactose, so they appear together and the results are showing the total glucose + galactose. These two residues (sum of glucose and galactose) were highly identified in all samples, followed by mannose and xylose (Figure 4).

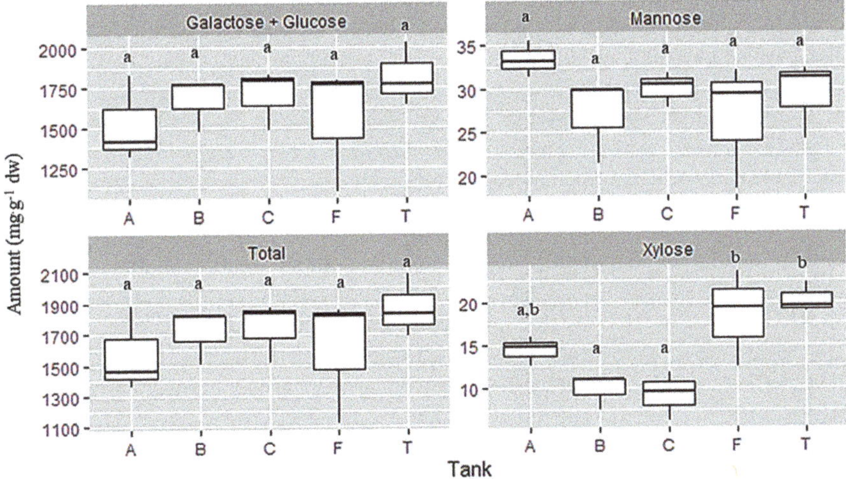

Figure 4. Total monosaccharide composition expressed in mg·g^{-1} dw from samples of *C. jubata* obtained from cultivated in indoor balloons (A, B and C), cultivated in outdoor tank (T) and collected from the field (F). The same upper letters (a, b) mean there are no differences between the treatments (cultivation type). Data as mean value, distribution and variance; $n = 3$.

3.3. Carrageenan Extraction

In the extraction of the main polysaccharide, all the samples have differences, with the F sample being the best sample assayed and the tank T cultivation (Outdoor) appearing to be the most stable sample in terms of extraction. In this case, all the indoor cultivations were inserted in only segment (Figure 5).

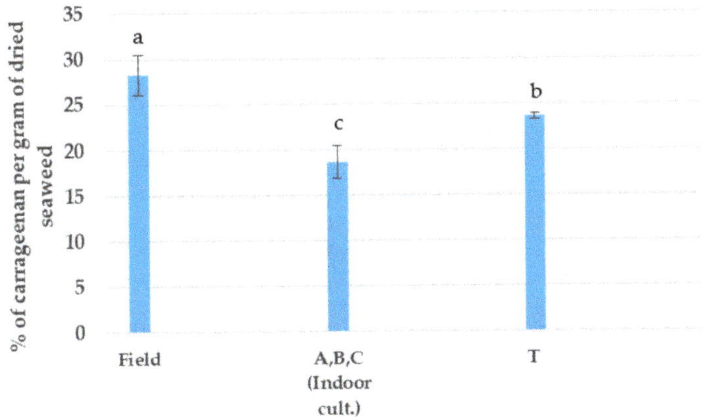

Figure 5. Yield of carrageenan per type of cultivation. [a, b, c] Equal letters indicate no significant differences at the *p*-value < 0.05 level.

4. Discussion

C. Jubata indoor cultivation has been studied for a long time by the research team, to obtain more information about this seaweed which has an interesting carrageenan, still with scarce or null information in literature. The cultivation in the lab was proven to be feasible to obtain more information about the performance of this seaweed to be cultivated. However, the inshore cultivation was not possible to be done in triplicated due to the lack of available cultivation tanks, although this first assay demonstrated interesting results in all the assays. This happened mainly due to the cultivation location being near the collection site and the water being considered identical by other researchers that monitor the water quality, in various locations near the collection and cultivation sites. However, the growth was poor when compared to indoor cultivation, this can be explained to the abiotic and biotic factors impact in the semi-controlled cultivation systems, the inshore cultivation system.

Samples collected in the field showed the highest value of fatty acids content since they were outdoors, where the conditions were not under control, due to the constant renewal of the water and the nutrients availability. After field samples, the higher value of fatty acid concentration was followed by tank T which was outside and in a larger proportion of cultivated seaweed. Indeed, the conditions in tank T were more similar to the field environmental factors.

In the carrageenan yield, the indoor culture was so identical in results that we put together to have more robust analysis. With that analysis, we observed that indoor cultivation presented lower content in carrageenan and more content in FAs than the inshore cultivation.

This is the first time that this species and genus is targeted to be characterized in the fatty acids and carbohydrates and there is a general lack of information about this seaweed, Whose cultivation is more feasible in lower temperatures [28]. However, this species is identical with high rated food consumed red seaweeds in the fatty acids and carbohydrates levels, as *Palmaria palmata*, *Chondrus crispus* and *Neopyropia tenera* [32,33].

The carrageenan extraction yield from all the samples is identical with the work of Pereira [21] in the same location of seaweed collection, however lower when compared to the work of the Zinoun and Cosson [28], that obtained 29% at the lowest yield of *C. jubata*, in which was collected in Cap Levy, in France.

Information about carbohydrates content is null in the literature, this being the first study characterizing the carbohydrate profile of this species and in *Calliblepharis* genus.

5. Conclusions

Our results support that seaweed indoor cultivation can help to optimize the overall seaweed cultivation allowing the assess to the ideal growth conditions of *C. Jubata*. However, this controlled cultivation system had a negative effect in the *C. Jubata* biochemical profile. Thus, the inshore cultivation (partially controlled system) showed more similar results with field samples than with the indoor cultivation system. However, the seaweed inshore cultivation needs to be more assayed to have high growth rate without lowering the biochemical compounds. The procedure of cultivation needs to be optimized to achieve and guarantee high content profiles and thus the best nutritional value.

Further and next plans of the study are to optimize and obtain more information about the aquaculture process of this seaweed species and the full characterization of the nutritional value of wild and cultivated *C. jubata* in parallel with the optimization procedure to guarantee to the stakeholders and farmers the best quality cultivation. The biochemical profile needs to be further analyzed in the quantification and characterization of the uronic acids and sulfate compounds, protein content and amino-acids profile, ash quantification and mineral concentration to have a useful tool to promote the *C. Jubata* cultivation.

Author Contributions: Conceptualization, G.S.A., J.C. and A.M.M.G.; Seaweed Cultivation, G.S.A., J.C. and T.M.; Fatty Acids and Carbohydrates analysis, A.L. and S.G.-P.; Carrageenan research, G.S.A. and J.C.; writing—original draft preparation, G.S.A., J.C., T.M., A.L. and S.G.-P.; writing—review and editing, J.C., A.M.M.G. and L.P.; supervision, A.M.M.G. and L.P.; project coordination, A.M.M.G.; funding acquisition, A.M.M.G. and L.P. All authors have read and agreed to the published version of the manuscript.

Funding: This work is financed by national funds through FCT - Foundation for Science and Technology, I.P., within the scope of the projects UIDB/04292/2020 – MARE - Marine and Environmental Sciences Centre and UIDP/50017/2020+UIDB/50017/2020 (by FCT/MTCES) granted to CESAM - Centre for Environmental and Marine Studies. This work was financed by the Live Food Production Laboratory (LABPAV) and the Tropical Aquaculture Study Group (GEAQUI) of the Federal Institute of Education, Science and Technology of Ceará—IFCE, Campus Aracati, Ceará, Brazil. João Cotas thanks to the European Regional Development Fund through the Interreg Atlantic Area Program, under the project NASPA (EAPA_451/2016). Adriana Leandro thanks FCT for the financial support provided through the doctoral grant SFRH/BD/143649/2019 funded by National Funds and Community Funds through FSE. Sara García-Poza thanks to the project MENU—Marine Macroalgae: Alternative recipes for a daily nutritional diet (FA_05_2017_011) which co-financed this research, funded by the Blue Fund under Public Notice No. 5—Blue Biotechnology. Ana M. M. Gonçalves acknowledges University of Coimbra for the contract IT057-18-7253.

Conflicts of Interest: The authors declare no conflict of interest.

References

1. Dawes, C.J. *Marine Botany*; Dawes, C.J., Ed.; John Wiley & Sons: New York, NY, USA, 1995; ISBN 978-0-471-19208-4.

2. Arasaki, S.; Arasaki, T. *Vegetables from the Sea*; Japan Publishing Inc.: Tokyo, Japan, 1983.

3. Wong, K.H.; Cheung, P.C.K. Nutritional evaluation of some subtropical red and green seaweeds. Part I—Proximate composition, amino acid profiles and some physico-chemical properties. *Food Chem.* **2000**, *71*, 475–482. [CrossRef]

4. Armisen, R. World-wide use and importance of Gracilaria. *J. Appl. Phycol.* **1995**, *7*, 231–243. [CrossRef]

5. Leandro, A.; Pacheco, D.; Cotas, J.; Marques, J.C.; Pereira, L.; Gonçalves, A.M.M. Seaweed's Bioactive Candidate Compounds to Food Industry and Global Food Security. *Life* **2020**, *10*, 140. [CrossRef] [PubMed]

6. Pereira, L. Biological and therapeutic properties of the seaweed polysaccharides. *Int. Biol. Rev.* **2018**, *2*, 1–50. [CrossRef]

7. Wijesekara, I.; Kim, S.K. Application of Marine Algae Derived Nutraceuticals in the Food Industry. *Mar. Algae Extr. Process. Prod. Appl.* **2015**, *2*, 627–638. [CrossRef]

8. De Jesus Raposo, M.F.; De Morais, A.M.B.; De Morais, R.M.S.C. Marine polysaccharides from algae with potential biomedical applications. *Mar. Drugs* **2015**, *13*, 2967–3028. [CrossRef] [PubMed]

9. Michalak, I.; Chojnacka, K. Algae as production systems of bioactive compounds. *Eng. Life Sci.* **2015**, *15*, 160–176. [CrossRef]

10. De Boer, J.A. *A Report on the Fisheries Training and Development Project*; FAO Fisheries and Aquaculture: Nassau, Bahamas, 1981.

11. Zhang, H.; Mao, W.; Fang, F.; Li, H.; Sun, H.; Chen, Y.; Qi, X. Chemical characteristics and anticoagulant activities of a sulfated polysaccharide and its fragments from Monostroma latissimum. *Carbohydr. Polym.* **2008**, *71*, 428–434. [CrossRef]

12. Farias, W.R.L.; Nazareth, R.A.; Mourão, P.A.S. Dual effects of sulfated D-galactans from the red algae Botryocladia occidentalis preventing thrombosis and inducing platelet aggregation. *Thromb. Haemost.* **2001**, *86*, 1540–1546. [CrossRef]

13. Zhou, G.; Sheng, W.; Yao, W.; Wang, C. Effect of low molecular λ-carrageenan from Chondrus ocellatus on antitumor H-22 activity of 5-Fu. *Pharmacol. Res.* **2006**, *53*, 129–134. [CrossRef]

14. Fu, Y.W.; Hou, W.Y.; Yeh, S.T.; Li, C.H.; Chen, J.C. The immunostimulatory effects of hot-water extract of Gelidium amansii via immersion, injection and dietary administrations on white shrimp Litopenaeus vannamei and its resistance against *Vibrio alginolyticus*. *Fish Shellfish Immunol.* **2007**, *22*, 673–685. [CrossRef] [PubMed]

15. Cotas, J.; Leandro, A.; Pacheco, D.; Gonçalves, A.M.M.; Pereira, L. A comprehensive review of the nutraceutical and therapeutic applications of red seaweeds (*Rhodophyta*). *Life* **2020**, *10*, 19. [CrossRef]

16. Ahmed, N.; Thompson, S. The blue dimensions of aquaculture: A global synthesis. *Sci. Total Environ.* **2019**, *652*, 851–861. [CrossRef] [PubMed]

17. Cardozo, K.H.M.; Guaratini, T.; Barros, M.P.; Falcão, V.R.; Tonon, A.P.; Lopes, N.P.; Campos, S.; Torres, M.A.; Souza, A.O.; Colepicolo, P.; et al. Metabolites from algae with economical impact. *Comp. Biochem. Physiol. C Toxicol. Pharmacol.* **2007**, *146*, 60–78. [CrossRef] [PubMed]

18. Buschmann, A.H.; Camus, C.; Infante, J.; Neori, A.; Israel, Á.; Hernández-González, M.C.; Pereda, S.V.; Gomez-Pinchetti, J.L.; Golberg, A.; Tadmor-Shalev, N.; et al. Seaweed production: Overview of the global state of exploitation, farming and emerging research activity. *Eur. J. Phycol.* **2017**, *52*, 391–406. [CrossRef]

19. García-Poza, S.; Leandro, A.; Cotas, C.; Cotas, J.; Marques, J.C.; Pereira, L.; Gonçalves, A.M.M. The Evolution Road of Seaweed Aquaculture: Cultivation Technologies and the Industry 4.0. *Int. J. Environ. Res. Public Health* **2020**, *17*, 6528. [CrossRef]

20. Allan, G.L.; Fielder, D.S.; Fitzsimmons, K.M.; Applebaum, S.L.; Raizada, S. *Inland Saline Aquaculture*; Woodhead Publishing Limited: Cambridge, UK, 2009; ISBN 9781845693848.

21. Pereira, L. Estudos em macroalgas carragenófitas (*Gigartinales, Rhodophyceae*) da costa portuguesa—Aspectos ecológicos, bioquímicos e citológicos. Ph.D. Thesis, University of Coimbra, Coimbra, Portugal, 2004.

22. Gaspar, R.; Fonseca, R.; Pereira, L. *Guia Ilustrado das Macroalgas da Baía de Buarcos, Figueira da Foz, Portugal*; Leonel Pereira: Coimbra, Portugal, 2020.

23. Zinoun, M.; Diouris, M.; Potin, P.; Floc'h, J.Y.; Deslandes, E. Evidence of Sulfohydrolase Activity in the Red Alga *Calliblepharis jubata*. *Bot. Mar.* **1997**, *40*, 49–54. [CrossRef]

24. Jouanneau, D.; Guibet, M.; Boulenguer, P.; Mazoyer, J.; Smietana, M.; Helbert, W. New insights into the structure of hybrid κ-/μ-carrageenan and its alkaline conversion. *Food Hydrocoll.* **2010**, *24*, 452–461. [CrossRef]

25. Deslandes, E.; Bodeau-Bellion, C.; Floc'h, J.Y.; Penot, M. 13C-NMR spectroscopy and chemical analysis of the carrageenans of four red algae (Gigartinales). *Plant Physiol. Biochem.* **1990**, *28*, 65–69.

26. Pereira, L.; Critchley, A.T.; Amado, A.M.; Ribeiro-Claro, P.J.A. A comparative analysis of phycocolloids produced by underutilized versus industrially utilized carrageenophytes (*Gigartinales, Rhodophyta*). *J. Appl. Phycol.* **2009**, *21*, 599–605. [CrossRef]

27. Pereira, L. *Edible Seaweeds of the World*; CRC Press: Boca Raton, FL, USA, 2016; ISBN 9780429154041.

28. Zinoun, M.; Cosson, J. Seasonal variation in growth and carrageenan content of Calliblepharis jubata (*Rhodophyceae, Gigartinales*) from the Normandy coast, France. *J. Appl. Phycol.* **1996**, *8*, 29–34. [CrossRef]

29. Gonçalves, A.M.M.; Pardal, M.Â.; Marques, S.C.; Mendes, S.; Fernández-Gómez, M.J.; Galindo-Villardón, M.P.; Azeiteiro, U.M. Responses of copepoda life-history stages to climatic variability in a southern-European temperate estuary. *Zool. Stud.* **2012**, *51*, 321–335.

30. Coimbra, M.A.; Waldron, K.W.; Selvendran, R.R. Isolation and characterisation of cell wall polymers from olive pulp (*Olea europaea* L.). *Carbohydr. Res.* **1994**, *252*, 245–262. [CrossRef]

31. Pereira, L.; van de Velde, F. Portuguese carrageenophytes: Carrageenan composition and geographic distribution of eight species (*Gigartinales, Rhodophyta*). *Carbohydr. Polym.* **2011**, *84*, 614–623. [CrossRef]

32. Pereira, L. A review of the nutrient composition of selected edible seaweeds. In *Seaweed: Ecology, Nutrient Composition and Medicinal Uses*; Pomin, V.H., Ed.; Nova Science Publishers, Inc.: Hauppauge, NY, USA, 2011; pp. 15–47, ISBN 978-1-61470-878-0.

33. Schmid, M.; Kraft, L.G.K.; van der Loos, L.M.; Kraft, G.T.; Virtue, P.; Nichols, P.D.; Hurd, C.L. Southern Australian seaweeds: A promising resource for omega-3 fatty acids. *Food Chem.* **2018**, *265*, 70–77. [CrossRef]

Article

A Novel Singleton Giant Phage Yong-XC31 Lytic to the *Pyropia* Pathogen *Vibrio mediterranei*

Lihua Xu [1,2,†], Dengfeng Li [1,*,†], Yigang Tong [3,*], Jing Fang [4], Rui Yang [1], Weinan Qin [1], Wei Lin [1], Lingtin Pan [1] and Wencai Liu [1]

1. Zhejiang Key Laboratory of Marine Biotechnology, Ningbo University, Ningbo 315832, China; xulihua202102@163.com (L.X.); yangrui@nbu.edu.cn (R.Y.); 1811091046@nbu.edu.cn (W.Q.); weilin0577@163.com (W.L.); 1911091090@nbu.edu.cn (L.P.); liuwencai103@163.com (W.L.)
2. Collage of Food and Pharmaceutical Sciences, Ningbo University, Ningbo 315832, China
3. School of Life Science and Technology, Beijing University of Chemical Technology, Beijing 100029, China
4. Ocean College of Qinzhou University, Qinzhou 535099, China; fj19891101@gmail.com
* Correspondence: lidengfeng@nbu.edu.cn (D.L.); tong.yigang@gmail.com (Y.T.); Tel.: +86-13819823176 (D.L.)
† These authors contributed equally to this work.

Citation: Xu, L.; Li, D.; Tong, Y.; Fang, J.; Yang, R.; Qin, W.; Lin, W.; Pan, L.; Liu, W. A Novel Singleton Giant Phage Yong-XC31 Lytic to the *Pyropia* Pathogen *Vibrio mediterranei*. *Appl. Sci.* **2021**, *11*, 1602. https://doi.org/10.3390/app11041602

Academic Editor: Marco F. L. Lemos

Received: 6 January 2021
Accepted: 5 February 2021
Published: 10 February 2021

Abstract: *Vibrio mediterranei* 117-T6 is extensively pathogenic to several *Pyropia* species, leading to the death of conchocelis. In this study, the first *V. mediterranei* phage (named *Vibrio* phage Yong-XC31, abbreviated as Yong-XC31) was isolated. Yong-XC31 is a giant phage containing an icosahedral head about 113 nm in diameter and a contractible tail about 219 nm in length. The latent period of Yong-XC31 is 30 min, and burst size is 64,227. Adsorption rate of Yong-XC31 to *V. mediterranei* 117-T6 can reach 93.8% in 2 min. The phage genome consisted of a linear, double-stranded 290,532 bp DNA molecule with a G + C content of 45.87%. Bioinformatic analyses predicted 318 open reading frames (ORFs), 80 of which had no similarity to protein sequences in current (26 January 2021) public databases. Yong-XC31 shared the highest pair-wise average nucleotide identity (ANI) value of 58.65% (below the ≥95% boundary to define a species) and the highest nucleotide sequence similarity of 11.71% (below the >50% boundary to define a genus) with the closest related phage. In the proteomic tree based on genome-wide sequence similarities, Yong-XC31 and three unclassified giant phages clustered in a monophyletic clade independently between the family Drexlerviridae and Herelleviridae. Results demonstrated Yong-XC31 as a new evolutionary lineage of phage. We propose a new phage family in Caudovirales order. This study provides new insights and fundamental data for the study and application of giant phages.

Keywords: *Vibrio mediterranei*; giant phage; complete genome

1. Introduction

Vibrios are ubiquitous in marine ecosystems living as well-described pathogens of aquatic fauna, for example, *Vibrio anguillarum*, *Vibrio harveyi*, *Vibrio coralliilyticus* and *Vibrio aestuarianus* [1–4]. *Vibrio mediterranei* is a potential emerging pathogen of marine animals such as corals and scallops [5,6]. *V. mediterranei* 117-T6 (CGMCC1.16311) was isolated from the bleached shell-born conchocelis of *Pyropia yezoensis* and was pathogenic to the conchocelis of several *Pyropia* species, including *P. yezoensis* [7,8]. *Pyropia* culture has a long history in China, but there are no effective prevention and control measures for bacterial diseases, which seriously affect the economic income of farmers and the healthy and sustainable development of *Pyropia* industry [8,9]. Global warming has caused an increase in sea surface temperature that has undoubtedly led to the unseasonal outbreaks of Vibrios as well as their increased abundance and virulence in marine environments and aquaculture [10,11]. Therefore, there are a growing need for effective methods for managing bacterial infections.

Viruses, in particular bacteriophages are the most abundant biological entities in the oceans, showing an extremely high, uncharted diversity [12]. The potential of viruses as therapeutic agents to treat infectious diseases has been known for a long time, and they are considered to be important agents and resources as a solution to antibiotic resistance [13]. The phages with double-stranded (ds) DNA genomes larger than 200 kbp are defined as giant or "jumbo" bacteriophages [14]. Giant bacteriophages commonly contain many genes that do not exist in the small genome bacteriophages. For example, some giant phages have several paralogous genes for DNA polymerase and RNA polymerase (RNAP). Importantly, the proteins encoded by these additional genes may replace the function of the host proteins, thereby reducing the dependence of giant phages on their bacterial hosts [15]. In addition, many genes in giant phages are interpreted as coding hypothetical proteins which are not found in small phages, and their biological features are understudied [14]. These genes may be a new resource of proteins for industrial, agricultural, or medical applications in the future [16–19]. However, large phages are not commonly isolated. Limited mobility on semi-solid plates, which prevented formation of visible plaques, may be the possible reason for the rare isolation of giant phage [15,20]. A centrifugal force that is too high, in the centrifugation to remove bacteria and protists, may also be a reason.

In recent years, bacteriophages have been increasingly applied for disease prevention and control in aquaculture [21–23]. To mine potential marine application resources for disease control of *Pyropia* vibriosis, the first *V. mediterranei* phage, named *Vibrio* phage Yong-XC31 (abbreviated as Yong-XC31), was isolated from the coastal water of Meishan island (29°46.989 N 121°57.516 E), Ningbo, China. Characteristics and the complete genome of the phage were analyzed. *Vibrio* phage Yong-XC31 is a giant phage presenting a new evolutionary lineage of phage.

2. Materials and Methods

2.1. Bacterial Strains and Culture Conditions

Vibrio mediterranei strain 117-T6 (China General Microbiological Culture Collection, CGMCC 1.16311) was provided by the Key Laboratory of Marine Biotechnology of Ningbo University [7,8]. *V. mediterranei* 117-T6 was cultured in NB seawater medium (peptone 10 g, beef extract 3 g, make final volume up to 1 L with filtered seawater, pH 7.2) at 29 °C with shaking at 180 rpm.

2.2. Antibiotic Susceptibility Test

Antibiotic susceptibility test was performed using antibiological susceptibility discs (Hangzhou Binhe Microorganism Reagent. Co., Ltd., Hangzhou, China), according to the instructions and following the standard of the Clinical & Laboratory Standards Institute (CLSI). 17 types of antibiotic disks were used as follows: penicillin G (6 ug), amoxicillin (10 μg), cephalexin (30 μg), kanamycin (30 μg), gentamicin (10 μg), tobramycin (10 μg), azithromycin (15 μg), aboren (30 μg), chloramphenicol (30 μg), ofloxacin (5 μg), tetracycline (30 μg), doxycycline (30 ug), rifampin (5 ug), trimethoprim/ sulphamethaxazole (1.25/23.75 μg), vancomycin (30 ug), polymyxin (30 ug) and clindamycin (2 ug). Fresh log phase *V. mediterranei* 117-T6 cultures were spread by sterial swab on Muller-Hinton agar medium, dried at room temperature (about 5 min). Antibiotic discs were then placed on the surface of the medium with triplicates. The plates were incubated at 29 °C for 16 to 18 h. The diameters of the inhibition zones formed were measured.

2.3. Phage Isolation and Morphological Observation

The surface seawater samples used for bacteriophage separation were collected from the seaside of Meishan island (29°46.989 N 121°57.516 E), Ningbo, China on 31 July 2018. The water samples were placed in an ice box and immediately brought back to the laboratory for treatment. After centrifugation at 10,000× *g* for 10 min, each 80 mL supernatant was mixed with 40 mL of 3 × NB seawater medium and 2 mL *V. mediterranei* 117-T6 of log phase (OD600 ≈ 0.6, 1.91 × 10^9 CFU/mL). The mixtures were cultured at 29 °C with shaking speed of 180 rpm for 3–4 h to enrich the phages, and then centrifuged at $10,000\times g$ for 10 min. The supernatants were filtered through 0.45 μm pore-size filters. Pure phage strain was obtained by three serial single-plaque isolation using the conventional double-layer agar method [24] employing *V. mediterranei* strain 117-T6 as the host. The bacteriophage was negatively stained with 3% uranium acetate for 20 s and observed under transmission electron microscope (Hitachi H-7650, Tokyo, Japan).

2.4. Thermal and pH Stability

Thermal stability was assessed by exposing aliquots of phage suspension (7.5×10^7 PFU/mL, 1 mL, in triplicates) to various temperatures (40, 50, 60 and 70 °C) for 30, 60, 90, 120, 150 and 180 min, respectively. To test the pH stability of Yong-XC31, aliquots of phage suspension (7.5×10^7 PFU/mL, 1 mL, in triplicates) were adjusted using NaOH or HCl to various pHs (2–12), withdrawn for 2 h at 29 °C. Titers of the treated and untreated control samples were measured using the double-layer plate method. The experiment was repeated three times.

2.5. MOI Selection Experiments and Adsorption Test

In multiplicities of infection (MOIs) selection experiments, 7.5×10^5 PFU of Yong-XC31 was mixed with a set of serial dilutions of 117-T6 cell suspensions (7.5×10^4 to 7.5×10^8 CFU) at MOIs of 0.001, 0.01, 0.1, 1 and 10, respectively, with triplicates. After 10 min of adsorption at 29 °C, the mixtures were centrifuged at $10,000\times g$ for 10 min. The precipitates were suspended in 5 mL NB seawater medium and incubated for 3 h at 29 °C with shaking speed of 180 rpm. Titers in the supernatant of the lysates were measured by using the double-layer agar plate method. The experiment was repeated three times. The MOI with the highest phage production was considered as the optimal one.

To evaluate optimum adsorption time, the phage was mixed with 117-T6 at the optimal MOI of 0.001 (phage-to-bacterium ratio = 7.5×10^5 PFU$/7.5 \times 10^8$ CFU) with triplicate and incubated at 29 °C with shaking speed of 180 rpm. Samples were taken at 0, 2, 4, 6, 8, 10, 15 and 20 min post-inoculation and centrifuged at $5000\times g$ for 10 min at 4 °C. Phage titers in the supernatant were measured by using the double-layer agar plate method. The experiment was repeated three times.

The influence of temperature and pH on the adsorption of the phage was monitored. The phage was mixed with 117-T6 at the optimal MOI of 0.001 as above and incubated 2 min at 0, 10, 20, 29, 40 and 45 °C, respectively, with triplicates, and then centrifuged at $5000\times g$ for 10 min at 4 °C. The phage was mixed with 117-T6 at the optimal MOI of 0.001 as above at 29 °C at the pH of 4, 5, 6, 7, 8 and 9, respectively, with triplicates, and then centrifuged at $5000\times g$ for 10 min at 4 °C. Phage titers in the supernatant were measured by using the double-layer agar plate method. The experiment was repeated three times.

2.6. One-Step Growth Experiment

Phage Yong-XC31 was mixed with 117-T6 (1.36×10^8 CFU/mL) at a MOI of 0.001 with triplicate and allowed to adsorb for 2 min at 29 °C. Then, centrifuged at $6000\times g$ for 10 min. Pellets containing the infected cells were washed twice and re-suspended in 40 mL of fresh NB seawater medium, incubated at 29 °C with shaking at 220 rpm. Samples were taken at 0, 10, 20, 30, 40, 50, 60, 90, 120, 150, 180 and 210 min, respectively, centrifuged at $5000\times g$ for 10 min at 4 °C, and then phage titers in the supernatant were immediately determined by the double-layer agar method. Titer measurements were conducted in triplicate.

2.7. Host Range Determination

The host range of Yong-XC31 was tested against 34 bacterial strains including the indicator host (Table 1) using the spot test according to the references [25,26]. Cultures of each bacterial strain (10^8 CFU/mL) were mixed with melted 0.7% agar (43 °C) NB seawater medium, respectively, with triplicates and poured on a 1.5% solid agar to make double layer agar plates. After solidification, 5 μL of phage suspensions (10^6 PFU/mL) were spotted on each plate. The plates were cultured at 29 °C for 8 h. It was considered as positive for clear lysis zones or plaques on the bacteria lawn.

Table 1. Bacterial strains used in the host range assay and infection results.

No. Bacteria Strains	Lytic Ability	No. Bacteria Strains	Lytic Ability
Vibrio mediterranei 117-T6	+	*Escherichia coli* DH5α	-
Vibrio hispanicus XXY	-	*Citrobacter freundii*	-
Vibrio fluvialis XXY1	-	*Shewanella putrefaciens*	-
Vibrio unlnificus XXY	-	*Streptococcus iniae*	-
Vibrio fluvialis XXY2	-	*Alternate pseudomonas* XY	-
Vibrio parahaemolyticus 1A08161	-	*Shigella dysenteriae*	-
Vibrio parahaemolyticus 1A11655	-	*Proteus vulgaris*	-
Vibrio alginolyticus LDF	-	*Proteus mirabilis*	-
Vibrio alginolyticus XY	-	*Aeromonas hydrophila*	-
Vibrio alginolyticus WY	-	*Shigella sonnei*	-
Vibrio alginolyticus SZT	-	*Pseudomonas aeruginosa* LDF	-
Vibrio harveyi 1–5	-	*Pseudomonas aeruginosa* SZT	-
Vibrio harveyi LDF	-	*Aeromonas sobria* LY-23	-
Vibrio pacinii XY	-	*Aeromonas sobria* ATCC43979	-
Vibrio anguillarum	-	*Pseudoalteromonas issachenkonii* XY	-
Aeromonas bivalvium XXY	-	*Salmonella paratyphi* B	-
Edwardsiella tarda SZT	-	*Enterobacter cloacae*	-
Edwardsiella tarda LW	-	*Marinomonas* sp.XY	-
Enterobacter sakazakii	-	*Pseudomonas* sp. XXY	-

(+) representative infection, (-) representative non-infection.

2.8. Genome Extraction and Sequencing

Genomic DNA of phage Yong-XC31 was extracted utilizing the modified method of standard phenol-chloroform extraction [27]. DNase I and RNase A (TransGen Biotech, Beijing, China) to a final concentration of 1 ug/mL were added to the purified phage Yong-XC31 stock solution. The mixture was incubated at 37 °C for 2 h to remove contaminating bacterial DNA and RNA. DNase I was deactivated by incubating the solution for 15 min at 80 °C. After adding EDTA to a final concentration of 20 mM, proteinase K at 50 ug/mL, and sodium dodecyl sulfate at 0.5% (*w/v*), the mixture was incubated for 1 h at 56 °C. An equal volume of phenol was added to extract the viral DNA, followed by centrifugation at $10,000 \times g$ for 10 min. The aqueous layer was moved to a fresh tube, to which an equal volume of phenol–chloroform-isoamyl alcohol (25:24:1) was added and mixed, and then centrifuged at $10,000 \times g$ for 10 min. The aqueous layer was mixed with isovolumic chloroform and centrifuged $10,000 \times g$ for 5 min. The aqueous phase was added with an equal volume of isopropanol, stored at −20 °C for 3 h, and centrifuged at 4 °C at $13,000 \times g$ for 20 min. The precipitated DNA was washed with 75% ethanol. The obtained Yong-XC31 DNA was resuspended in deionized water and stored at −20 °C for further experiments.

The NEBNext Ultra II DNA Library Prep Kit for Illumina (#E7645) was used for constructing genomic library. Genome sequencing of the phage was performed using the Illumina MiSeq (SanDiego, CA, USA) sequencing platform to obtain 2 × 300 bp paired-end reads. Low-quality (Q-value < 20) reads and adapters were filtered out using fastp. SPAdes 3.13.0 software (http://cab.spbu.ru/software/spades/) was utilized to assemble the trimmed reads. Phage genome termini were analyzed using our proposed method [28].

2.9. Genome Annotation and Taxonomic Analyses

ORFs prediction was initially conducted with RAST (http://www.rast.nmpdr.org), and then identified with HMMER and HHpred web server [29,30]. All predicted ORFs were manually verified using the BLAST tool of NCBI (https:/blast.ncbi.nlm.nih.gov/). tRNAscan-SE was used to search for tRNA genes (http://lowelab.ucsc.edu/tRNAscan-SE/; [31]), and the antibiotic resistance genes and virulence factors were searched in CARD database (http://arpcard.mcmaster.ca) and VFDB database (http://www.mgc.ac.Cn/VFs/main.htm), respectively.

Nucleotide sequence comparisons were firstly conducted using NCBI BLASTn [32]. The pair-wise average nucleotide identity (ANI) values between the giant phage Yong-XC31 and the phages with the 10 top highest homology (Table 2) in BLASTn comparison (e-value < 10^{-5}) were confirmed using OrthoANI [33]. The in silico DNA–DNA hybridization (isDDH) values between Yong-XC31 and these closely related giant phages were performed using GGDC (formula 2) [34]. The percentage of conserved proteins (POCP) values between Yong-XC31 and these related giant phages were calculated as described previously [35]. Genome comparison of Yong-XC31 and the closest related phage BONAISHI was done by a genome comparison visualizer, Easyfig [36]. To estimate the nucleotide sequence similarity between Yong-XC31 and other phages in current (26 January 2021) public databases, the Pairwise Sequence Comparison (PASC) classification tool [37] (https://www.ncbi.nlm.nih.gov/sutils/pasc/viridty.cgi) was used. ViPTree online [38] (available at https://www.genome.jp/viptree/) was used to generate a proteomic tree based on genome wide sequence similarities, computed by tBLASTx, gathering 56 classified phages of the nine families of Caudovirales, Yong-XC31 and 10 giant phages in Table 2.

Table 2. Average nucleotide identity (ANI), in silico DNA–DNA hybridization (isDDH) and percentage of conserved proteins (POCP) values between *Vibrio* phage Yong-XC31 and the giant phages with the 10 top highest homology in BLASTn comparison (e-value < 10^{-5}).

Strain	Accession no.	ANI, %	isDDH, %	POCP, %
Vibrio phage BONAISHI	MH595538	58.65	0	6.774
Aeromonas phage phiAS5	HM452126	0	0	0.606
Aeromonas phage CC2	JX123262	0	0	0.268
Aeromonas phage AS-yj	MF498774	0	0	0.278
Aeromonas phage AS-szw	MF498773	0	0	0.273
Aeromonas phage AS-zj	MF448340	0	0	0.272
Aeromonas phage AS-sw	MF498775	0	0	0.271
Pseudomonas phage Phabio	MF042360	0	0	0.26
Vibrio phage 2.275.O._10N.286.54.E11	MG592671	0	0	0.245
Vibrio phage 2 TSL-2019	MF063068	0	0	0

2.10. Genome Sequence Accession Number

The complete genome sequence of *Vibrio* phage Yong-XC31 is available from GenBank under nucleotide accession number MK308674. The phage has been deposited in China General Microbiological Culture Collection Center under number CGMCC No. 17098.

3. Results

3.1. Antibiotic Susceptibility of V. mediterranei 117-T6

V. mediterranei 117-T6 was resistant to 4 of 17 tested antibiotics, which were penicillin G, aboren, polymyxin and clindamycin, respectively (Figure 1). It was intermediate to azithromycin, vancomycin and gentamicin. Though, causing inhibition zones on *V. mediterranei* 117-T6 lawn, amoxicillin, rifampicin, tobramycin, tetracycline, doxycycline, chloramphenicol, ofloxacin and selectrin were not effective inhibitors to *V. mediterranei* 117-T6 as bacterial colonies existing in the inhibition zone of these antibiotics. Bacterium resistant to three or more antimicrobials was defined as multidrug resistant (MDR) [39,40]. *V. mediterranei* 117-T6 is a typical multidrug resistant bacterium.

3.2. Phage Morphology

Vibrio phage Yong-XC31 produced transparent circular plaques with clear and regular edges on *V. mediterranei* lawns (Figure 2a). Yong-XC31 is large in size having a head with icosahedral approximately spherical structure, about 113 nm in diameter, and a contractible tail, about 219 nm in length (Figure 2b).

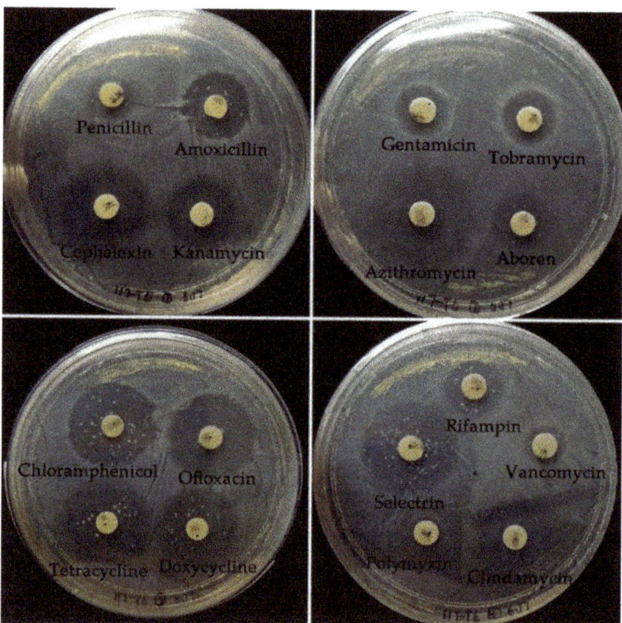

Figure 1. Results of antibiotic susceptibility test. The size of the inhibition zone indicates the susceptibility of bacteria to antibiotics following the standard of the Clinical & Laboratory Standards Institute (CLSI).

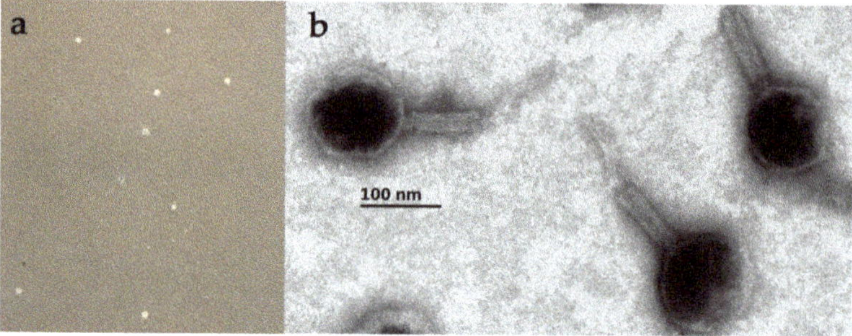

Figure 2. Morphology of the plaque and *Vibrio* phage Yong-XC31. (**a**) plaque produced by Yong-XC31 on *V. mediterranei* lawn; (**b**) electron micrograph of Yong-XC31. Bar represents 100 nm.

3.3. Thermal and pH Stability

Temperature and pH stability provide more latent capacity with regard to phage storage, transport and potential applications. In the thermal and pH stability assays, Yong-XC31 was very stable at 40 °C maintaining constant titer for over 3 h, relatively stable at 50 °C and 60 °C (Figure 3a). The phage was stable at pH 5 to 8. Although the titer declined dramatically at pH 2, 10 and 11 (Figure 3b), Yong-XC31 can tolerate a pH ranging from 2–4 and 9–10 for at least two hours without complete loss of infectivity.

3.4. Optimal MOI and Factors Influencing Adsorption

Among all the tested MOIs, the optimal MOI is 0.001, when mixing 7.5×10^5 PFU of Yong-XC31 with 7.5×10^8 CFU 117-T6 host cells (Figure 4a).

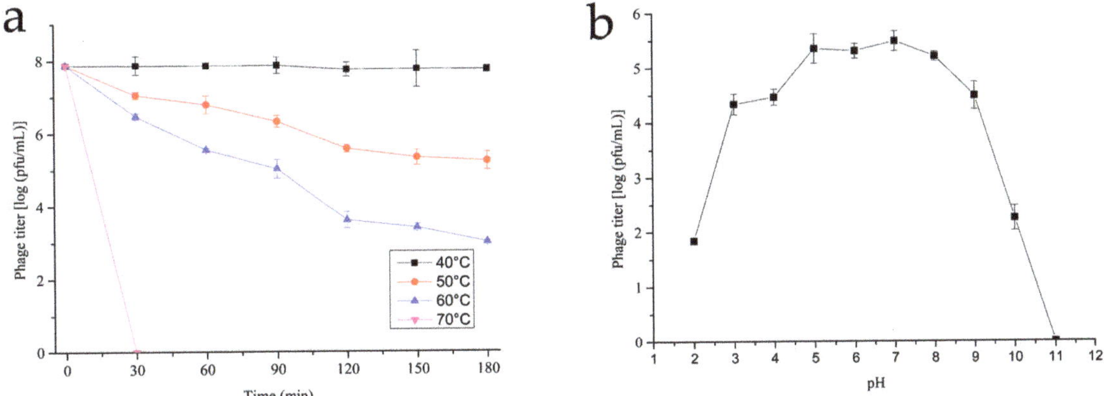

Figure 3. Thermostability (**a**) and pH stability (**b**) of *Vibrio* phage Yong-XC31. All the experiments were conducted in triplicates.

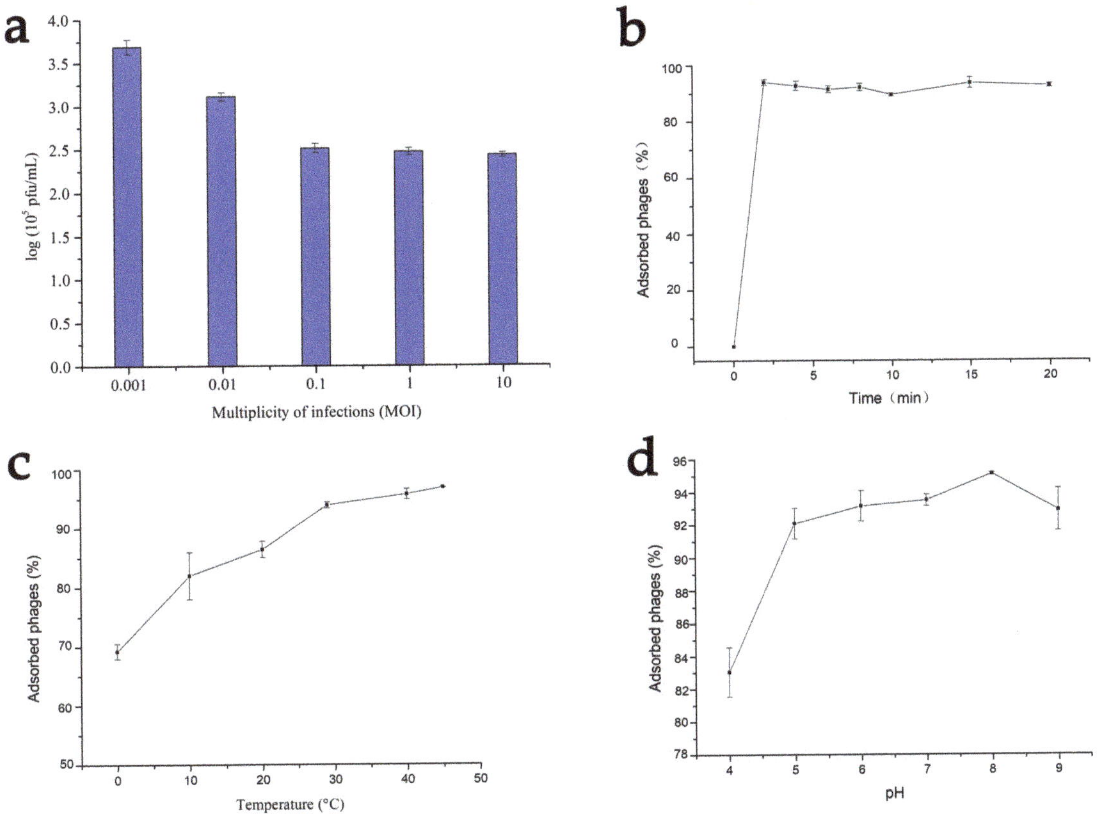

Figure 4. Multiplicities of infection (MOIs) curve (**a**), temporary adsorption kinetics of *Vibrio* phage Yong-XC31 (**b**), influence of temperature on adsorption kinetics (**c**), and influence of pH on adsorption kinetics (**d**). All the experiments were conducted with triplicates.

Adsorption is a key stage in virus recognition of a sensitive host cell. Adsorption of Yong-XC31 to *V. mediterranei* 117-T6 is very efficient. The adsorption rate can reach 93.8% in 2 min at 29 °C (Figure 4b). External conditions such as temperature and pH are influential to phage adsorption, in turn, affect potential applications. From 0 to 50 °C, the adsorption rate of Yong-XC31 to *V. mediterranei* 117-T6 increases with the increase of temperature (Figure 4c). From pH 4–8, the adsorption rate increases with the increase of pH, while decreased at pH 9 (Figure 4d).

3.5. One-Step Growth Curve

The one-step growth curve of Yong-XC31 showed a latent period of about 30 min. There are two generally accepted methods for calculating phage burst size. The average burst size of Yong-XC31 was 64,227 calculated as the ratio of mean yield of phage particles liberated to the mean phage particles that infected the bacteria in the latent period (Figure 5) referring to reference [41], and 2 PFU/cell calculated as the ration of the final count of liberated phage particles to the initial count of infected bacterial cells at the beginning of the latent period (Figure 5) referring to reference [24].

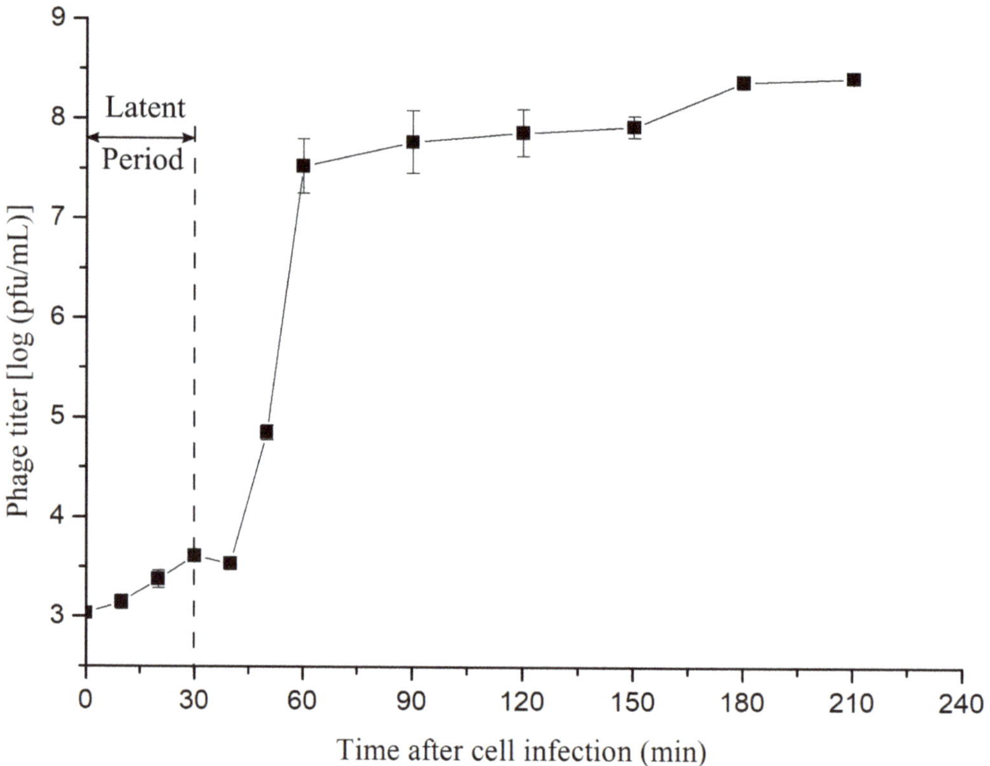

Figure 5. One-step growth curve. All the experiments were conducted at MOI = 0.001 with triplicates.

3.6. Host Range Determination

To test Yong-XC31 host range, 34 different bacterial strains (Table 1) were used employing spot test method. Among these bacterial strains, only *V. mediterranei* 117-T6 was found to be susceptible to phage Yong-XC31. Yong-XC31 showed strict host specificity, which may be favorable for the application of the phage as it is difficult to change the normal flora.

3.7. Genome Analysis of Yong-XC31

Yong-XC31 is a typical giant phage at genome level. Complete genome sequencing was conducted using next-generation sequencing (NGS) with an average read length of 284.6 bases. 81.66% of the sequencing reads were matched to the complete genome (133,281 out of 163,212 reads) with an average sequencing depth of 130.6-fold. The complete genome of Yong-XC31 is 290,532 bp long with a G + C content of 45.87%. Yong-XC31 genome is unique sharing the highest BLASTn homology with the most related phage BONAISHI (sequence ID: MH595538, query cover as low as 1%, identity 70.62%). 25.2% predicted Yong-XC31 ORFs might encode novel proteins as they have no similarity to protein sequences in current public databases (26 January 2021).

No tRNA gene was found in the genome. Among the 318 ORFs of Yong-XC31, 17.3% were predicted functions, 82.7% predicted as hypothetical proteins. No ORFs associated with virulence factors, toxins or antibiotic resistance genes were found among the annotated ones within Yong-XC31 genome. The predicted ORFs could be classified into five functional categories, including DNA replication/regulation, bacteriophage packaging, bacteriophage structure, lysis and hypothetical protein (Figure 6, Supplement Table S1).

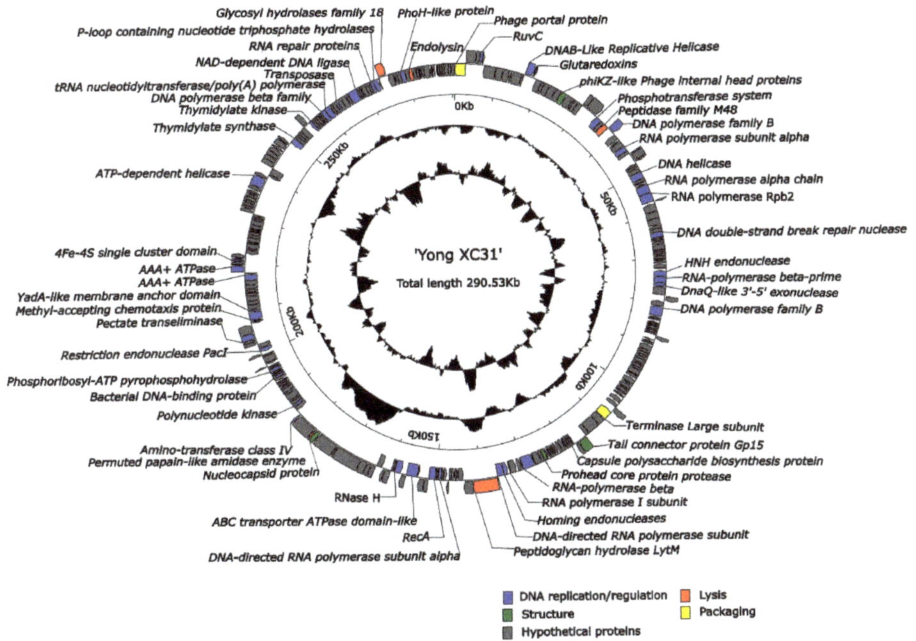

Figure 6. Genome map of *Vibrio* phage Yong-XC31. The arrow indicates the coding protein sequence, the protein function is distinguished by color, red stands for lysis, yellow for packaging, blue for DNA replication/regulation and green for structure, gray indicates hypothetical protein.

6 ORFs were annotated as structural and assembly genes (Supplement Table S1). 15 ORFs were annotated to be involved in DNA replication, recombination, and repair. 2 ORFs encode homing endonucleases of HNH family (ORF68 and ORF141), one of which is an intron-encoded homing endonuclease (ORF141) located between genes encoding subunits of RNAP.

As mentioned in the introduction, giant bacteriophages commonly contain many genes that do not exist in the small genome bacteriophages. Notably, Yong-XC31 harbours multiple enzyme genes related to bacteriolysis. At least 5 ORFs predicted to play a role in host cell lysis, including a peptidase (family M48, ORF29), a permuted papain-like amidase enzyme (ORF175), a glycosyl hydrolases (family 18, ORF295), a peptidoglycan hydrolase LytM (ORF143) and a lysin (ORF302). Remarkably, the lysin has a modular structure comprising an N-terminal glycoside hydrolase 19 domain, and a C-terminal peptidoglycan binding domain (PGBD).

As introduction described, some giant phages contain additional genes which may replace the function of the host proteins, thereby reducing the dependence of giant phages on their bacterial host. Noticeably, Yong-XC31 harbours eight paralogous genes (ORF34, ORF46, ORF47, ORF69, ORF136, ORF140, ORF142, ORF153) for RNAP. In particular, Yong-XC31 harbours genes (ORF265 and ORF271) encoding thymidylate synthase (dTMP synthase) and thymidylate kinase (TdR kinase) responsible for the de novo biosynthesis of thymidylate (dTMP) and as the salvage enzyme which leads to the production of dTMP even in presence of dTMP synthase inhibition [42]. Yong-XC31 also harbours a gene (ORF278) encoding tRNA nucleotidyltransferase/poly(A) polymerase participating translation, ribosomal structure and biogenesis [43].

3.8. Taxonomy

Yong-XC31 is a novel phage. As described above, Yong-XC31 shared the highest BLASTn identity (70.62%) with the closest related phage BONAISHI, with the query cover as low as 1%. Only 8 genes were found to be conserved in Yong-XC31 and *Vibrio* phage BONAISHI by the comparison of Yong-XC31 and *Vibrio* phage BONAISHI (Figure 7).

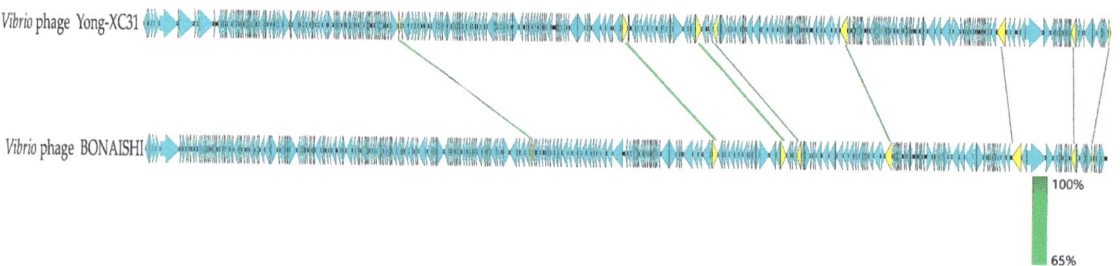

Figure 7. Genome comparison of *Vibrio* phage Yong-XC31 and the most closely related phage BONAISHI. Strain-specific protein coding genes are shown in light blue, protein coding genes conserved in Yong-XC31 and *Vibrio* phage BONAISHI are shown in yellow.

Further, the genome of Yong-XC31 shared only 0–58.65% ANI values with the giant phages with the 10 top highest homology in BLASTn comparison (e-value $< 10^{-5}$) (Table 2), which are substantially below the $\geq$95% ANI boundaries to define a species [44,45]. The isDDH values between Yong-XC31 and these giant phages all were 0, lower than the 70% cut off to define a species [34]. The POCP values between Yong-XC31 and giant phage species were 0.26–6.77%, much lower than the genus boundary cut-off of 50%.

In addition, the Bacterial and Archaeal Viruses Subcommittee (BAVS) within the International Committee on the Taxonomy of Viruses (ICTV), which holds the responsibility of classifying new prokaryotic viruses, recently redefined a genus as a cohesive group of viruses sharing a >50% of high degree of nucleotide sequence similarity [45,46]. Yong-XC31 shared the highest nucleotide sequence similarity, as low as 11.71%, with the closest related phage in PASC [37] search. It's much below the >50% boundaries to define a genus. Summarily, these results indicate the status of Yong-XC31 as a taxonomically unique phage presenting a novel monophyletic genus.

Most jumbo phages were attributed to the Myoviridae family. Yet, in the proteomic tree constructed based on genome wide sequence similarities (Figure 8), Yong-XC31 and three unclassified giant phages clustered in a monophyletic clade independently between the family Drexlerviridae and Herelleviridae. These four giant phages harbor genomes of 242,446–309,157 bp. A new phage family is proposed in Caudovirales order.

Figure 8. The proteomic tree is generated using ViPTree online based on genome-wide similarities determined by tBLASTx. 56 type species of nine families belonging to the order Caudovirales, *Vibrio* phage Yong-XC31, and 10 giant phages in Table 2 were included in the analysis. Bacteriophage family assignments according to the official ICTV classification are provided with different color bars. No color bar is marked to unclassified giant phages. Red star refers to Yong-XC31. The sizes or size ranges of the giant phage genomes are shown in red font on the right.

4. Discussion

Pyropia is an important cultivated seaweed in Northeast Asia [7,47]. In recent years, increase in the prevalence of diseases and pests has caused a subsequent reduction in their quantity and commercial value [47,48]. Among these diseases, yellow spot disease (YSD) is a destructive disease of the conchocelis sporeling culture of *Pyropia* [7,47,48]. *V. mediterranei* 117-T6 (CGMCC1.16311) was pathogenic to the YSD of conchocelis of several

Pyropia species [8]. Traditional antibiotic therapy has no stable effect on YSD [49]. In this study, antibiotic susceptibility test demonstrated that *V. mediterranei* 117-T6 was multidrug resistant, which accord with the ineffective antibiotic treatment. Therefore, the development of new tools and strategies to control pathogens and treat diseased *Pyropia* is becoming a most important issue.

Phages are currently emerging as potential treatments for multidrug resistant bacterial infections. Yet, literature search yielded no information about *V. mediterranei* phage. In this study, the first *V. mediterranei* phage was isolated and identified as a novel giant phage representing a novel phage genus. Proteomic tree indicates that Yong-XC31 together with giant *Vibrio* phage BONAISHI, *Vibrio* phage 2 TSL-2019 and *Pseudomonas* phage Phabio represent a new evolutionary lineage independently between the family Drexlerviridae and Herelleviridae. The four giant phages all infect Gram-negative bacteria, contain genome of 242,446–309,157 bp with only an average 1/4 of the predicted genes being assigned functions. A new family is proposed here in Caudovirales. In the novel proposed family, Yong-XC31 presents a monophyletic genus in consideration of the pair-wise ANI, isDDH, POCP and PASC values.

The therapeutic value of a candidate bacteriophage relies on the characterization of viral properties such as stability, growth kinetics, host range and viral yield [50]. Results of this study indicated that the temperature and pH stability of Yong-XC31 was good, which is conducive to its application to control *V. mediterranei* infections in complex environment. Results also demonstrated other good properties of Yong-XC31 including a very low optimal MOI (0.001), efficient adsorption to host (reach 93.8% adsorption rate in only 2 min), a short latent period (30 min), a high burst size (64,227) and strict host specificity. Additionally, our small-scale laboratory experiments proved that Yong-XC31 could protect both the free-living conchocelis (FLC) and the shell-borne conchocelis (SBC) of *Pyropia* from harm caused by *V. mediterranei* [49]. These results suggest significant application potential of Yong-XC31 in *Pyropia* production.

Thus, far, only fourteen jumbo phages infecting *Vibrio* bacteria have been characterized. Among them, Yong-XC31 is most related to the unclassified phage BONAISHI, which is strictly lytic for several strains of *V. coralliilyticus* pathogenic to coral [50]. Yong-XC31 and BONAISHI share several special characteristics. Their genomes are both about of 290 kb, contain more than one paralogous gene encoding RNAPs and do not harbour identified tRNA gene.

Most phages rely on RNAPs of a bacterial host to transcribe their genes [51]. A strategy used by some phages is to depend on their own single-subunit RNAPs to transcribe a subset of viral genes [52]. An even more radical strategy is used by some giant phage, not need to rely on the host RNAP, yet only relies on self-encoded phage multi-subunit RNAPs: virion-associated RNAPs and early expressed RNAPs [14]. These RNAPs can perform two functions. The virion-associated RNAP is injected into the host cell together with phage DNA and transcribes early phage genes [53]. The early expressed RNAP formed by putative RNAP subunits which would transcribe viral genes expressed in the middle and late stages of infection [54]. This study finds that Yong-XC31 contains five genes encoding virion-associated RNAPs (ORF34, ORF136, ORF140, ORF142, ORF153) and three early expressed RNAPs (ORF46, ORF47, ORF69). This may be beneficial to its efficient infection. In addition, the absence of detectable tRNA in phage BONAISHI genome suggests that it is well adapted to the translation machinery of its hosts, which is a critical process for efficient phage propagation [50]. Similarly, we did not find any tRNA in the genome of Yong-XC31.

As mentioned above, alternative therapies must be developed to mitigate the sharp increase in antibiotic resistance. Besides phage itself, novel antimicrobial strategies include enzyme-based antibiotics ("enzybiotics") such as phage lytic enzymes. Yong-XC31 harbors multiple (at least eight) enzyme genes related to bacteriolysis. These enzyme genes may be potential beneficial resources.

Very little is known about the processes of host–phage interaction in marine environments. Bailey et al., found that *Synechococcus* phage S-PM2 contain genes encode homologs of the key photosystem II reaction center core polypeptides (D1 and D2) and proposed that this might play an active role in protecting their hosts against photoinhibition, thereby ensuring an energy supply for replication by preventing the deleterious effects on host cell integrity seen during acute photoinhibition [55]. Phages infecting marine picocyanobacteria often carry a *psbA* gene, which encodes a homolog to the photosynthetic reaction center protein, D1. Bragg et al., proposed that phage encoded D1 may help to maintain photosynthesis during the lytic cycle, which in turn could bolster the production of deoxynucleoside triphosphates (dNTPs) for phage genome replication [56]. They examined the contribution of phage *psbA* expression to phage genome replication under constant low irradiance and predicted that phage *psbA* expression could lead to an increase in the number of phage genomes produced during a lytic cycle of between 2.5 and 4.5% [56]. In this study, Yong-XC31 possesses unusual genes rarely present in other phages. In particular, Yong-XC31 contains an ORF encoding Methyl-accepting chemotaxis protein (MCP) and a gene encoding EIIB belonging phosphotransferase system. As the predominant chemoreceptor and signal transducer in bacteria and archaea, MCPs also termed transducer-like proteins (Tlps), serve as sensors in bacterial chemotactic signaling [57]. MCPs sense intracellular and environmental signals, and relay them to the downstream signaling pathways in the cytoplasm. Then, bacteria utilize the well-known chemotactic responses to move towards factors that favor survival [58]. Further studies are needed to find whether the expression of the viral methyl-accepting chemotaxis protein gene may enhance the host survival and persistence in the complex environment, thus profiting the phage itself. The phosphoenol phosphotransferase system (PTS) is a multi-component signal transduction cascade that regulates diverse aspects of bacterial cellular physiology in response to the availability of high-energy sugars in the environment [59]. In bacteria, there are often many different EIIs. EIIs are responsible for the phosphorylation of the carbohydrate as well as its transportation across the bacterial membrane [60]. The significance of gene encoding EIIB in Yong-XC31 remains to be studied.

5. Conclusions

In conclusion, *Vibrio* phage Yong-XC31 was isolated and characterized as a new potential biocontrol strategy to control *Vibrio mediterranei*. A total of 80 orphan ORFs in Yong-XC31 may be a new resource of applications. Giant phage Yong-XC31 has very unique genome sequence. Blast search results, ANI values, isDDH values, POCP values, nucleotide sequence similarity estimated via PASC and proteomic tree demonstrated Yong-XC31 as a singleton phage distantly related to previously sequenced bacteriophages, representing a new evolutionary lineage of phage. We propose a new phage family with Yong-XC31 as the representative specie of a genus. This study expands the diversity of phages.

6. Patents

Lihua Xu; Dengfeng Li; Jing Fang et al., A virulent phage vB_VmeM-Yong XC31 of *Vibrio mediterranei* and its application (ZL201910792955.8).

Supplementary Materials: The following are available online at https://www.mdpi.com/2076-341 7/11/4/1602/s1.

Author Contributions: D.L., L.X. and Y.T. designed the research. L.X., D.L., Y.T., J.F., R.Y., W.Q., W.L. (Wei Lin), L.P. and W.L. (Wencai Liu) performed the research. L.X. & D.L. analyzed data and wrote the paper. All authors have read and agreed to the published version of the manuscript.

Funding: This research was funded by the national key research and development program (2018YFA 0903000); the open fund of key laboratory of marine biogenetic resources of State Oceanic Administration (HY201602), the Public Welfare Science and Technology Plan of Ningbo (202002N3039) and sponsored by K.C. Wong Magna Fund of Ningbo University; National Natural Science Foundation of China (31772871).

Institutional Review Board Statement: Ethical review and approval were not applicable for studies not involving humans or animals.

Informed Consent Statement: Not applicable for studies not involving humans.

Data Availability Statement: Data is contained within the article or supplementary material.

Acknowledgments: We are very grateful for high quality technical support provided by PingPing Zhan of the Electron Microscopy Laboratory of Ningbo University.

Conflicts of Interest: The authors declare no conflict of interest. All authors have seen the manuscript and approved to submit to your journal. To the best of our knowledge, this manuscript has not been published in whole or in part nor is it being considered for publication elsewhere.

References

1. Ben-Haim, Y.; Rosenberg, E. *Vibrio coralliilyticus* sp. nov., a temperature-dependent pathogen of the coral *Pocillopora damicornis*. *Int. J. Syst. Evol. Microbiol.* **2003**, *53*, 309–315. [CrossRef]
2. Pang, L.; Zhang, X.H.; Zhong, Y.; Chen, J.; Li, Y.; Austin, B. Identification of *Vibrio harveyi* using PCR amplification of the *toxR* gene. *Lett. Appl. Microbiol.* **2006**, *43*, 249–255. [CrossRef] [PubMed]
3. Romalde, J.L.; Dieguez, A.L.; Lasa, A.; Balboa, S. New *Vibrio* species associated to molluscan microbiota: A review. *Front. Microbiol.* **2014**, *4*, e413. [CrossRef] [PubMed]
4. Goudenège, D.; Travers, M.A.; Lemire, A.; Petton, B.; Haffner, P.; Labreuche, Y.; Tourbiez, D.; Mangenot, S. A single regulatory gene is sufficient to alter *Vibrio aestuarianus* pathogenicity in oysters. *Environ. Microbiol.* **2015**, *17*, 4189–4199. [CrossRef] [PubMed]
5. Rubio-Portillo, E.; Yarza, P.; Penalver, C.; Ramos-Esplá, A.A.; Anton, J. New insights into *Oculina patagonica* coral diseases and their associated *Vibrio* spp. communities. *ISME J.* **2014**, *8*, 1794–1807. [CrossRef] [PubMed]
6. Serrano, W.; Tarazona, U.I.; Olaechea, R.M.; Friedrich, M.W. Draft genome sequence of a new *Vibrio* strain with the potential to produce bacteriocin-like inhibitory substances, isolated from the gut microflora of scallop (*Argopecten purpuratus*). *Genome Announc.* **2018**, *6*, e00419-18. [CrossRef]
7. Liu, Q.Q.; Xu, M.Y.; He, Y.Y.; Tao, Z.; Yang, R.; Chen, H.M. Complete genome sequence of *Vibrio mediterranei* 117-T6, a potentially pathogenic bacterium isolated from the conchocelis of *Pyropia* spp. *Microbiol. Resour. Announc.* **2019**, *8*, e01569-01518. [CrossRef] [PubMed]
8. Xu, M.Y.; Yang, R.; Liu, Q.Q.; He, Y.Y.; Chen, H.M. Study on yellow spot disease in *Pyropia* species infected by *Vibrio mediterranei* 117-T6. *J. Fish. China* **2020**, *44*, 1–11. [CrossRef]
9. Wang, X.L.; He, L.W.; Ma, Y.C.; Huan, L.; Wang, Y.Q.; Xia, B.M.; Wang, G.C. Economically important red algae resources along the Chinese coast: History, status, and prospects for their utilization. *Algal Res.* **2020**, *46*, 101817. [CrossRef]
10. Baker-Austin, C.; Trinanes, J.A.; Taylor, N.G.H.; Hartnell, R.; Siitonen, A.; Martinez-Urtaza, J. Emerging *Vibrio* risk at high latitudes in response to ocean warming. *Nat. Clim. Chang.* **2013**, *3*, 73–77. [CrossRef]
11. Vezzulli, L.; Grande, C.; Reid, P.C.; Hélaouët, P.; Edwards, M.; Höfle, M.G.; Brettar, I.; Colwell, R.R.; Pruzzo, C. Climate infuence on *Vibrio* and associated human diseases during the past half-century in the coastal North Atlantic. *Proc. Natl. Acad. Sci. USA* **2016**, *113*, E5062–E5071. [CrossRef]
12. Suttle, C.A. Marine viruses—major players in the global ecosystem. *Nat. Rev. Microbio.* **2007**, *5*, 801–812. [CrossRef] [PubMed]
13. Breitbart, M.; Bonnain, C.; Malki, K.; Sawaya, N.A. Phage puppet masters of the marine microbial realm. *Nat. Microbiol.* **2018**, *3*, 754–766. [CrossRef] [PubMed]
14. Yuan, Y.H.; Gao, M.Y. Jumbo bacteriophages: An overview. *Front. Microbiol.* **2017**, *8*, 403. [CrossRef]
15. Hendrix, R.W. Jumbo bacteriophages. *Curr. Top. Microbiol. Immunol.* **2009**, *328*, 229–240. [CrossRef]
16. Gorski, A.; Miedzybrodzki, R.; Łobocka, M.; Glowackarutkowska, A.; Bednarek, A.; Borysowski, J.; Jonczykmatysiak, E.; Łusiakszelachowska, M.; Weberdąbrowska, B.; Baginska, N.J.V. Phage therapy: What have we learned? *Viruses* **2018**, *10*, 288. [CrossRef] [PubMed]
17. Moye, Z.D.; Woolston, J.; Sulakvelidze, A. Bacteriophage applications for food production and processing. *Viruses* **2018**, *10*, 205. [CrossRef]
18. Dy, R.L.; Rigano, L.A.; Fineran, P.C. Phage-based biocontrol strategies and their application in agriculture and aquaculture. *Biochem. Soc. Trans.* **2018**, *46*, 1605–1613. [CrossRef] [PubMed]
19. Jamal, M.; Bukhari, S.M.A.U.S.; Andleeb, S.; Ali, M.; Raza, S.; Nawaz, M.A.; Hussain, T.; Rahman, S.U.; Shah, S.S.A. Bacteriophages: An overview of the control strategies against multiple bacterial infections in different fields. *J. Basic Microbiol.* **2019**, *59*, 123–133. [CrossRef]
20. Van Etten, J.L.; Lane, L.C.; Dunigan, D.D. DNA viruses: The really big ones (Giruses). *Annu. Rev. Microbiol.* **2010**, *64*, 83–99. [CrossRef]
21. Wang, Y.H.; Barton, M.; Elliott, L.; Li, X.X.; Abraham, S.; O'Dea, M.; Munro, J. Bacteriophage therapy for the control of *Vibrio harveyi* in greenlip abalone (*Haliotis laevigata*). *Aquaculture* **2017**, *473*, 251–258. [CrossRef]

22. Prada-Peñaranda, C.; Salazar, M.; Güiza, L.; Pérez, M.I.; Leidy, C.; Vives-Florez, M.J. Phage preparation FBL1 prevents *Bacillus licheniformis* biofilm, bacterium responsible for the mortality of the Pacific white shrimp *Litopenaeus vannamei*. *Aquaculture* **2018**, *484*, 160–167. [CrossRef]

23. Jun, J.W.; Kim, J.H.; Shin, S.P.; Han, J.E.; Chai, J.Y.; Park, S.C. Protective effects of the *Aeromonas* phages pAh1-C and pAh6-C against mass mortality of the cyprinid loach (*Misgurnus anguillicaudatus*) caused by *Aeromonas hydrophila*. *Aquaculture* **2013**, *416*, 289–295. [CrossRef]

24. Zhang, Q.; Xing, S.Z.; Sun, Q.; Pei, G.Q.; Cheng, S.; Liu, Y.Y.; An, X.P.; Zhang, X.L.L.; Qu, Y.G.; Tong, Y.G. Characterization and complete genome sequence analysis of a novel virulent Siphoviridae phage against *Staphylococcus aureus* isolated from bovine mastitis in Xinjiang, China. *Virus Genes* **2017**, *53*, 464–476. [CrossRef] [PubMed]

25. Nikapitiya, C.; Dananjaya, S.H.S.; Chandrarathna, H.P.S.U.; Senevirathne, A.; De Zoysa, M.; Lee, J. Isolation and characterization of multidrug resistance *Aeromonas salmonicida* subsp. *Salmonicida* and its infecting novel phage ASP-1 from goldfish (*Carassius auratus*). *Indian J. Microbiol.* **2019**, *59*, 161–170. [CrossRef]

26. Adams, M.H. *Bacteriophages*; Interscience Publishers Inc.: New York, NY, USA; London, UK, 1959.

27. Lu, S.G.; Le, S.; Tan, Y.L.; Zhu, J.M.; Li, M.; Rao, X.C.; Zou, L.Y.; Li, S.; Wang, J.; Jin, X.L.; et al. Genomic and proteomic analyses of the terminally redundant genome of the *Pseudomonas aeruginosa* phage PaP1: Establishment of genus PaP1-like phages. *PLoS ONE* **2013**, *8*, e62933. [CrossRef]

28. Zhang, X.L.L.; Wang, Y.H.; Li, S.S.; An, X.P.; Pei, G.Q.; Huang, Y.; Fan, H.; Mi, Z.Q.; Zhang, Z.Y.; Wang, W.; et al. A novel termini analysis theory using HTS data alone for the identification of *Enterococcus* phage EF4-like genome termini. *BMC Genom.* **2015**, *16*, 1–11. [CrossRef]

29. Finn, R.D.; Clements, J.; Eddy, S.R. HMMER web server: Interactive sequence similarity searching. *Nucleic Acids Res.* **2011**, *39*, W29–W37. [CrossRef]

30. Söding, J.; Biegert, A.; Lupas, A.N. The HHpred interactive server for protein homology detection and structure prediction. *Nucleic Acids Res.* **2005**, *33*, W244–W248. [CrossRef]

31. Lowe, T.M.; Eddy, S.R. tRNAscan-SE: A program for improved detection of transfer RNA genes in genomic sequence. *Nucleic Acids Res.* **1997**, *25*, 955–964. [CrossRef]

32. Johnson, M.; Zaretskaya, I.; Raytselis, Y.; Merezhuk, Y.; McGinnis, S.; Madden, T.L. NCBI BLAST: A better web interface. *Nucleic Acids Res.* **2008**, *36*, W5–W9. [CrossRef] [PubMed]

33. Lee, I.; Kim, Y.O.; Park, S.C.; Chun, J. OrthoANI: An improved algorithm and software for calculating average nucleotide identity. *Int. J. Syst. Evol. Microbiol.* **2016**, *66*, 1100–1103. [CrossRef] [PubMed]

34. Meier-Kolthoff, J.P.; Auch, A.F.; Klenk, H.P.; Göker, M. Genome sequence-based species delimitation with confidence intervals and improved distance functions. *BMC Bioinform.* **2013**, *14*, 60. [CrossRef] [PubMed]

35. Qin, Q.L.; Xie, B.B.; Zhang, X.Y.; Chen, X.L.; Zhou, B.C.; Zhou, J.; Oren, A.; Zhang, Y.Z. A proposed genus boundary for the prokaryotes based on genomic insights. *J. Bacteriol.* **2014**, *196*, 2210–2215. [CrossRef]

36. Sullivan, M.J.; Petty, N.K.; Beatson, S.A. Easyfig: A genome comparison visualizer. *Bioinformatics* **2011**, *27*, 1009–1010. [CrossRef]

37. Bao, Y.; Chetvernin, V.; Tatusova, T. Improvements to pairwise sequence comparison (PASC): A genome-based web tool for virus classification. *Arch. Virol.* **2014**, *159*, 3293–3304. [CrossRef]

38. Nishimura, Y.; Yoshida, T.; Kuronishi, M.; Uehara, H.; Ogata, H.; Goto, S. ViPTree: The viral proteomic tree server. *Bioinformatics* **2017**, *33*, 2379–2380. [CrossRef] [PubMed]

39. Magiorakos, A.P.; Srinivasan, A.; Carey, R.B.; Carmeli, Y.; Falagas, M.E.; Giske, C.G.; Harbarth, S.; Hindler, J.F.; Kahlmeter, G.; Olsson-Liljequist, B.; et al. Multidrug-resistant, extensively drug-resistant and pandrug-resistant bacteria: An international expert proposal for interim standard definitions for acquired resistance. *Clin. Microbiol. Infect.* **2012**, *18*, 268–281. [CrossRef]

40. Mupfunya, C.R.; Qekwana, D.N.; Naidoo, V. Antimicrobial use practices and resistance in indicator bacteria in communal cattle in the Mnisi community, Mpumalanga, South Africa. *Vet. Med. Sci.* **2020**. [CrossRef]

41. Walakira, J.K.; Carrias, A.A.; Hossain, M.J.; Jones, E.; Terhune, J.S.; Liles, M.R. Identification and characterization of bacteriophages specific to the catfish pathogen, *Edwardsiella ictaluri*. *J. Appl. Microbiol.* **2008**, *105*, 2133–2142. [CrossRef] [PubMed]

42. Look, K.Y.; Moore, D.H.; Sutton, G.P.; Prajda, N.; Abonyi, M.; Weber, G. Increased thymidine kinase and thymidylate synthase activities in human epithelial ovarian carcinoma. *Anticancer Res.* **1997**, *17*, 2353–2356.

43. Reuven, N.B.; Zhou, Z.; Deutscher, M.P. Functional overlap of tRNA nucleotidyltransferase, poly(A) polymerase I, and polynucleotide phosphorylase. *J. Biol. Chem.* **1997**, *272*, 33255–33259. [CrossRef] [PubMed]

44. Richter, M.; Rosselló-Móra, R. Shifting the genomic gold standard for the prokaryotic species definition. *Proc. Natl. Acad. Sci. USA* **2009**, *106*, 19126–19131. [CrossRef] [PubMed]

45. Adriaenssens, E.M.; Brister, J.R. How to name and classify your phage: An informal guide. *Viruses* **2017**, *9*, 70. [CrossRef] [PubMed]

46. Adriaenssens, E.M.; Sullivan, M.B.; Knezevic, P.; van Zyl, L.J.; Sarkar, B.L.; Dutilh, B.E.; Alfenas-Zerbini, P.; Łobocka, M.; Tong, Y.G.; Brister, J.R.; et al. Taxonomy of prokaryotic viruses: 2018-2019 update from the ICTV Bacterial and Archaeal Viruses Subcommittee. *Arch. Virol.* **2020**, *165*, 1253–1260. [CrossRef]

47. Yang, R.; Liu, Q.Q.; He, Y.Y.; Tao, Z.; Xu, M.Y.; Luo, Q.J.; Chen, J.J.; Chen, H.M. Isolation and identification of *Vibrio mediterranei* 117-T6 as a pathogen associated with yellow spot disease of *Pyropia* (Bangiales, Rhodophyta). *Aquaculture* **2020**, *526*, 735372. [CrossRef]

48. Ward, G.M.; Faisan, J.P., Jr.; Cottier-Cook, E.J.; Gachon, C.; Hurtado, A.Q.; Lim, P.E.; Matoju, I.; iMsuya, F.E.; Bass, D.; Brodie, J. A review of reported seaweed diseases and pests in aquaculture in Asia. *J. World Aquac. Soc.* **2020**, *51*, 815–828. [CrossRef]

49. Zhu, J.K.; Xu, M.Y.; Liu, Q.Q.; Li, D.F.; Yang, R.; Chen, H.M. Bacteriophage therapy on the conchocelis of *Pyropia haitanensis* (Rhodophyta) infected by *Vibrio mediterranei* 117-T6. *Aquaculture* **2021**, *531*, 735853. [CrossRef]

50. Jacquemot, L.; Bettarel, Y.; Monjol, J.; Corre, E.; Halary, S.; Desnues, C.; Bouvier, T.; Ferrier-Pagès, C.; Baudoux, A.C. Therapeutic potential of a new jumbo phage that infects *Vibrio coralliilyticus*, a widespread coral pathogen. *Front. Microbiol.* **2018**, *9*, 2501. [CrossRef]

51. Lavysh, D.; Sokolova, M.; Minakhin, L.; Yakunina, M.; Artamonova, T.; Kozyavkin, S.; Makarova, K.S.; Koonin, E.V.; Severinov, K. The genome of AR9, a giant transducing *Bacillus* phage encoding two multisubunit RNA polymerases. *Virology* **2016**, *495*, 185–196. [CrossRef]

52. Ceyssens, P.J.; Minakhin, L.; Van den Bossche, A.; Yakunina, M.; Klimuk, E.; Blasdel, B.; De Smet, J.; Noben, J.P.; Bläsi, U.; Severinov, K.; et al. Development of giant bacteriophage φKZ is independent of the host transcription apparatus. *J. Virol.* **2014**, *88*, 10501–10510. [CrossRef]

53. Thomas, J.A.; Weintraub, S.T.; Wu, W.; Winkler, D.C.; Cheng, N.; Steven, A.C.; Black, L.W. Extensive proteolysis of head and inner body proteins by a morphogenetic protease in the giant *Pseudomonas aeruginosa* phage φKZ. *Mol. Microbiol.* **2012**, *84*, 324–339. [CrossRef]

54. Yakunina, M.; Artamonova, T.; Borukhov, S.; Makarova, K.S.; Severinov, K.; Minakhin, L. A non-canonical multisubunit RNA polymerase encoded by a giant bacteriophage. *Nucleic Acids Res.* **2015**, *43*, 10411–20104. [CrossRef] [PubMed]

55. Bailey, S.; Clokie, M.R.J.; Millard, A. Cyanophage infection and photoinhibition in marine cyanobacteria. *Res. Microbiol.* **2004**, *155*, 720–725. [CrossRef] [PubMed]

56. Bragg, J.G.; Chisholm, S.W. Modeling the fitness consequences of a Cyanophage-encoded photosynthesis gene. *PLoS ONE* **2008**, *3*, e3550. [CrossRef] [PubMed]

57. Salah Ud-Din, A.I.M.; Roujeinikova, A. Methyl-accepting chemotaxis proteins: A core sensing element in prokaryotes and archaea. *Cell. Mol. Life Sci.* **2017**, *74*, 3293–3303. [CrossRef] [PubMed]

58. He, K.; Bauer, C.E. Chemosensory signaling systems that control bacterial survival. *Trends Microbiol.* **2014**, *22*, 389–398. [CrossRef]

59. Pickering, B.S.; Lopilat, J.E.; Smith, D.R.; Watnick, P.I. The transcription factor Mlc promotes *Vibrio cholerae* biofilm formation through repression of phosphotransferase system components. *J. Bacteriol.* **2014**, *196*, 2423–2430. [CrossRef]

60. Shin, D.H. A preliminary X-ray study of a refolded PTS EIIB^{fruc} protein from *Escherichia coli*. *Protein Pept. Lett.* **2008**, *15*, 630–632. [CrossRef] [PubMed]

MDPI

St. Alban-Anlage 66

4052 Basel

Switzerland

Tel. +41 61 683 77 34

Fax +41 61 302 89 18

www.mdpi.com

Applied Sciences Editorial Office

E-mail: applsci@mdpi.com

www.mdpi.com/journal/applsci